# The Body as Matter in Art and Science

Karolina Żyniewicz

# The Body as Matter in Art and Science

## An Autoethnography of Transmattering

palgrave
macmillan

Karolina Żyniewicz
Berlin, Berlin, Germany

ISBN 978-3-032-13664-0 ISBN 978-3-032-13665-7 (eBook)
https://doi.org/10.1007/978-3-032-13665-7



This Palgrave Macmillan imprint is published by the registered company Springer Nature Switzerland AG.
The registered company address is: Gewerbestrasse 11, 6330 Cham, Switzerland

*For all human and non-human beings accompanying my daily transmatterive activities.*

# Acknowledgments

I would like to express my deepest gratitude to Professor Paweł Golik and Professor Szymon Wróbel for their invaluable patience, feedback, and courage in becoming liminal together with me. Also, I could not have made this journey without all the biologists who generously hosted me in their laboratories and were my partners and supporters in developing the two liminal projects that form the core of this book:

At the Faculty of Biology, University of Warsaw:

Professor Anna Maria Ciemerych-Litvinienko
Dr Jakub Piątkowski
Dr Agata Kodroń
Dr Magdalena Kaliszewska
Dr Agnieszka Piotrowska-Nowak
Dr Karolina Łabędzka-Dmoch
Dr Bohdan Paterczyk
Dr Radosław Stachowiak

At the Nencki Institute of Experimental Biology:

Professor Ewa Sikora
Dr Anna Bielak-Żmijewska
Dr Wioleta Grabowska-Pyrzewicz
Dr Agnieszka Bojko

At the Instituto de Medicina Molecular, Lisbon:

Professor Luís Graça

I would also like thank to all the biologists whom I have met on other occasions than working on the projects discussed in this book. Exchanges with them were essential for my research and for my personal development. It would be a sheer impossibility to name all these individuals here, so I will only mention: Professor Wojciech Branicki and Dr Magdalena Kukla-Bartoszek (Malopolska Centre of Biotechnology, Jagiellonian University), and Professor Paweł Stepień and Professor Katarzyna Tońska (Institute of Genetics and Biotechnology, Faculty of Biology, University of Warsaw).

*Competing Interests* The author has no competing interests to declare that are relevant to the content of this manuscript.

# Contents

# About the Author

**Karolina Żyniewicz** (PL/DE) is a Berlin-based artist&researcher&educator, a liminal being, existing and performing between various contexts and disciplines. She is a graduate of the Faculty of Visual Arts of the Strzemiński Academy of Fine Arts in Łódź and holds a PhD in Cultural Sciences (obtained in the transdisciplinary program Nature-Culture at Artes Liberales Faculty, University of Warsaw). While executing art&research projects based mainly on biotechnology and medicine, she also conducts ethnographic and autoethnographic observations. In her work, she underlines the epistemic and didactic dimensions of art. In 2016–2018, she cooperated with the education departments of the Museum of Modern Art in Warsaw and the Zachęta National Gallery of Art, running workshops and museum lessons for regular audiences and with accessibility for visually impaired and deaf visitors. She uses these educational experiences in her own liminal practice. The topics of her work are often related to broadly understood life and death and their social and biological dimensions. That is why she is a member of the Institute of the Good Death in Warsaw. Her work is strongly process-oriented, so she often participates in residencies, like the European Media Art Platform (EMAP) at Kontejner Zagreb or Coalesce in Buffalo, NY, and group projects, like Stretching Senses School by Humboldt Cluster of Excellence Matters of Activity. She has exhibited internationally, for instance, at Kunstquartier Bethanien in Berlin, Kontejner Zagreb, or The Center of Contemporary Art Łaźnia in Gdańsk, Poland. Additionally, she is an associate reviewer of the *Technoetic Arts journal.*

# List of Figures

CHAPTER 1

# Introduction: On Matter, Mattering, and Transmattering

This book is dedicated to matter, mattering, and transmattering. Matter, understood as a biological substance, meaning, and importance, is the main protagonist of my narrative. It is crucial that I am the matter to transmatter in and through my research—my body and its elements, my social and cognitive position, my ideas and entanglements. I asked myself several questions: How can I use myself as research matter? What can I discover while transmattering myself? What kind of story can be plotted by the matter of my body? This is a self-reflective analysis of a series of experiments that I have done to answer these questions.

Throughout this text, I will call myself a liminal being to characterize my cognitive position of being 'in-between': between various creative fields, including academia, art, and science (biology). My primary field is art, but I decided to leave my comfort zone to check what was happening outside, in humanities departments and biological laboratories. The starting point of the cognitive journeys I decided to analyze and describe was artistic concepts of two projects: *safe suicide* and *The Last Supper*. These concepts required using scientific methods in the laboratory environment. From the very beginning, they were thought of as fieldwork circumstances, providing a framework for autoethnographic reflection. All the fields listed above worked as contexts transmattering my personal matter and, at the same time, my personal matter had the function of transmattering these contexts.



K. Żyniewicz, *The Body as Matter in Art and Science*,
https://doi.org/10.1007/978-3-032-13665-7_1

Crucially, my personal matter does not work alone. It is in constant interactions and entanglements with other forms of matter, other (human and non-human) bodies, other meanings, and other priorities. This enmeshment makes any transmatterive operation very challenging and complex, and it also makes the analysis and description of these processes complicated and utterly intriguing.

Matter in my research is an ontological condition—everything that we need to know to feel situated in the world. As analysing the entirety of existing matter and its aspects would, of course, exceed the capacity of this book, I decided to limit my endeavour to considering my personal matter and all its elements during the realization of the *safe suicide* and *The Last Supper* projects.

Matter is an ontological notion, and mattering is an epistemological operation conducted on matter. Many scholars have already done mattering. At a certain moment in my empirical work, ideas developed by, mostly feminist, New Materialism scholars appeared as elements of matter to which I needed to relate in creating my relational networks. The theoretical concepts proposed by Donna Haraway, Karen Barad, Maria Puig de la Bellacasa, Serpil Oppermann, Serenella Iovino, and Jane Bennett helped me to construct my own story about matter. In my vision, mattering describes and analyses matter in its transformations, living, and dying.

Being my invention, the term 'transmattering' stems from my fieldwork experiences and from building my own mattering story (Fig. 1.1). Transmattering as such moves beyond theoretical mattering through liminal practice that combines art, science, technology, and the humanities. According to Barad, physical matter and its meaning cannot exist separately. My goal was to see and describe what their correlation looked like in the case of my hands-on transmatterive experiences. I needed to analyse my personal matter and its entanglement, make sense of that, and investigate whether transforming my personal matter (and its network) automatically changed its meaning and, if so, how.

The role of this text is to guide the reader through theoretically depicted mattering and empirically experienced transmattering. Transmattering is my research, which approaches matter from the viewpoint of my embodiment, a viewpoint of a liminal human being. I show how my liminal practice enabled me to transgress the mattering vision I derived from the literature and from the natural cycle of living and dying. The main question that I decided to pose was: How can one transmatter? The text relates how my autoethnographic research answered this question. I present one

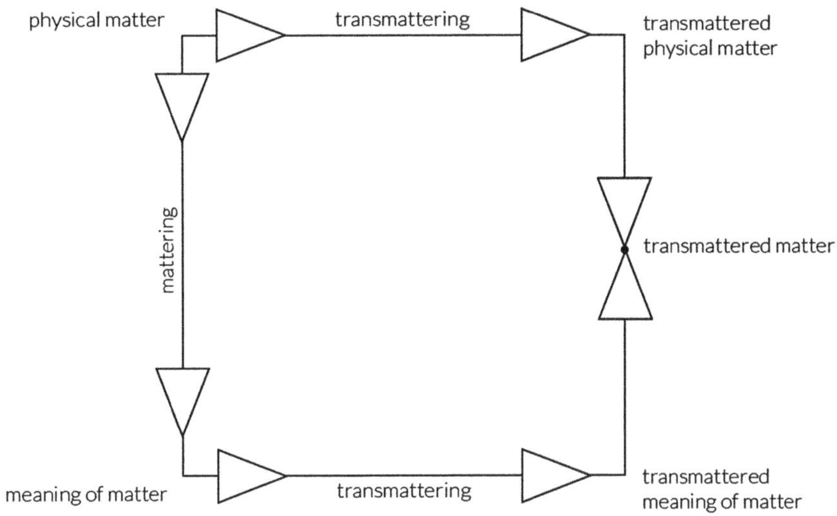


**Fig. 1.1** Mattering and transmattering: a scheme of operations

of many possible ways to transmatter, which nobody can in fact repeat in the same form. The knowledge produced in my research and my writing is supposed to be mattered to be transmattered by others in their own hands-on experiences.

The prefix ‘trans’ indicates that I processed available theoretical and biological matter and selected matter in a specific way that enabled me to be bodily present in my research. The process is described and analysed without pretending that I offer an objective point of view.

‘Trans’ is also related to transgression understood in the Foucauldian sense as the constant crossing of borders without removing them. According to Michel Foucault:

> Transgression is an action which involves the limit, that narrow zone of a line where it displays the flash of its passage, but perhaps also its entire trajectory, even its origin; it is likely that transgression has its entire space in the line it crosses. The play of limits and transgression seems to be regulated by a simple obstinacy: transgression incessantly crosses and recrosses a line which closes up behind it in a wave of extremely short duration, and thus it is made to return once more right to the horizon of the uncrossable. But this relationship is considerably more complex: these elements are situated

> in an uncertain context, in certainties which are immediately upset so that thought is ineffectual as soon as it attempts to seize them. (1977: 33–4)

I understand boundaries as properties of matter produced in the mattering processes. As Foucault suggests, one can imagine boundaries as lines to cross or narrow border zones, where we can experience the differentiation of one form of matter from another. The experience can involve physical, sensual, or intellectual operations related to naming and describing. The boundaries of matter are not rigid or static; they are in constant movement and reconfiguration. In my research, I consider a variety of borders distinguished by language: physical borders, social borders, and borders between various disciplines of cultural production. My research work is not about crossing these borders but about performing/vibrating in the frontier area. In my study, the border zone is related to liminality, which I understand as a state where I am focused more on transforming than on precisely classifying matter. The stability of the border zone is a condition of my existence, allowing me to transmatter matter from three different fields.

This book is not conceived as a mere report on my trajectory; rather, it is dynamic matter, co-created with many human and non-human agents. Using personal matter and its entanglement makes this research unique. Readers are invited to use their knowledge and experiences to transmatter what I wrote and created in my transmattering operations.

The feminist takes on of materialism provided theoretical underpinnings for my research by offering various ways of mattering. The most important thing was paying attention to the complexity of matter and relations of various material elements (including my personal matter). Haraway's becoming-with (2016), which echoed Susan Merrill Squier's becoming liminal (2004), stressed the importance of interspecies kinship and the collectivity of existence. These ideas helped me to capture the character of my personal matter and its entanglement with other material agencies in the three fields I explored. The idea of intra-actions, which I borrowed from Barad's ground-breaking work (2007), added another crucial aspect to my analysis of intra-material relations. Barad's concept rejects the notion of interacting, yet separate material agencies and replaces it with the idea of joint agency resulting from intra-actions of material agents. In the light of this idea, entanglements of various forms of matter are fundamental to all phenomena that occur. Puig de la Bellacasa augmented my vision of relations between my personal matter and other forms of matter with the importance of care. Her perspective lent legitimacy to emotional and vulnerable experiences of researchers, including myself. Analysing Bellacasa's understanding of care, I realized that it was crucial to my transmatterive activity in all its varieties. Care about my personal matter, especially the matter extracted from my body to become a research sample and an art medium, and care about the project ideas and all human and non-human beings intra-acting with me added up to the primary method of my work and also formed the foundation of my professional ethics. Besides analysing how material elements were interrelated, I also had to pay attention to their behaviour and characteristics. For that purpose, I adopted Bennett's vitalism, which emphasizes that all material pieces are active. They vibrate. This idea led me to define myself as a material entanglement constantly co-vibrating with other forms of vibrant matter. Vibration is an action that does not require passing any distance or, in my case, crossing borders physically. I could stay in the liminal/frontier zone and co-vibrate with an array of material elements in the three different fields and contexts.

Bennett's vision of entanglement worked in my mind as a bridge between feminist materialist ideas and empirically grounded Science and Technology Studies (STS) theories. Bennett's concept of assemblages as a mode of the self-organization of matter will be explained in the following chapters, yet it is important to stress that I saw a balance between Barad's intra-actions, Bruno Latour's Actor–Network Theory (ANT), and Bennett's assemblages. Latour was one of my three major sources on Science and Technology Studies (STS) and the ethnography of the laboratory, the other two inspirations being Karin Knorr Cetina's concept of epistemic cultures and Robert Mitchell's definition of laboratory artefacts. All these concepts worked as theoretical and methodological matter to transmatter. New Materialist visions focused on material reality in general, while STS research was performed on its specific section: laboratories. It was essential for my empirical study to understand how the laboratory had been described by researchers of cognitive processes. Having grasped that, I developed my vision of the laboratory from a liminal practitioner's perspective.

This book is a material story to tell. I follow Oppermann's and Iovino's contributions to the New Materialist basket of ideas. They have put forward the concept of "storied matter", where various forms of matter are not only endowed with agency; they also wave stories (Oppermann 2018: 411). Human bodies are not only active participants of storied matter; they also work as tools for articulating, writing, and sharing material stories. In my research, my body plays both these roles when it is entangled in storied matter and when it writes about this material journey.

Besides explaining my personal matter and its intra-acts, I had to characterize my cognitive and social position or dynamic. I am not static. I am in constant vibration, intra-acting with the matter of art, academia, and the biological laboratory. The border (or frontier) zone is also a liminal zone, a space of vibrating without either crossing the borders around for good or removing them (Fig. 1.2). The anthropological concepts proposed by Arnold van Gennep and Victor Turner helped me to understand liminality in general and in my particular case. I define my position in my research and in my text as a liminal position that determines my relations with other human and non-human agents, including the readers of this text. In my study, the liminal zone is a convenient transmattering site. However, continual transmatterive vibrations can be exhausting and sometimes risky in the clash with the regime of disciplinary division.

The liminal situation called for a suitable methodology that made it possible to analyse the liminal condition of my personal matter and its intra-actions with matter located beyond the borders. Autoethnography proved to be the only viable opportunity not only for building a narrative from the perspective of my entangled personal matter but also for

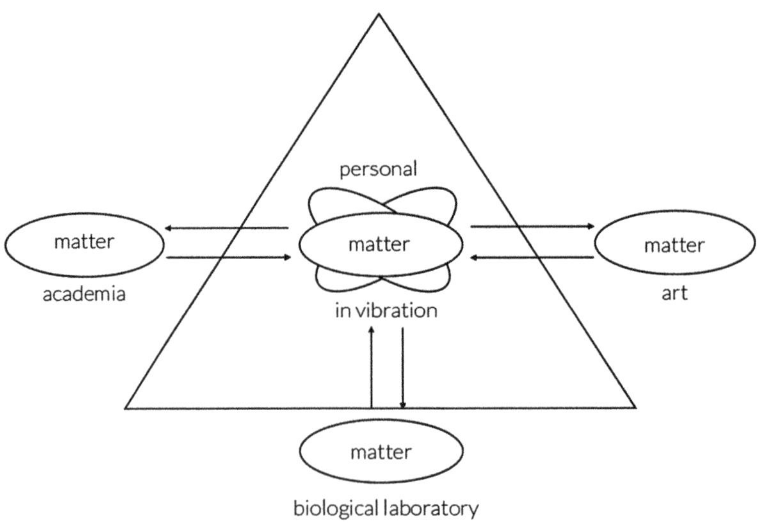


**Fig. 1.2** Transmatterive intra-actions

conveying the performativity of transmatterive vibrating. Autoethnography allowed me to incorporate emotionality and vulnerability as essential aspects of care-informed research.

While I call my projects *safe suicide* and *The Last Supper* liminal (happening thanks to transmatterive vibrations), they are as a rule officially classified as biological art (bioart). However, from the start, I have understood my transmatterive projects as a form of evocative autoethnography, founded on artistic (visual/aural/interactive/symbolic) language and at the same time on academic, discursive language. The location/vibration "in-between" has helped me to send different messages to each field and simultaneously to remain sensitive to responses coming from them.

The two artistic projects involved different transformative operations, although both were connected by association to my personal matter. They underwent a transmatterive translation from the scientific context to the artistic one at a certain moment. That is why it was crucial to depict the laboratory environment and then the gallery setting. In crafting an overview of the laboratory circumstances, I paid attention to my relations with the human inhabitants of this zone. I intended to follow social and semantic transmattering when, being myself a liminal being as an artist and a humanities researcher, I entered a biological laboratory to conduct liminal projects based on laboratory methods. In this book, I provide an account of the trajectory of my transmatterive work in several biological laboratories, describing how the laboratory work looks from a liminal practitioner's perspective. My account also explains how scientists and laboratory workers react to a liminal practitioner's presence and how they understand the sense of a liminal project conducted in their environment. The general description of what happens when a liminal practitioner enters a biological laboratory also works as a transmattering setup. These considerations are followed by an analysis of biological/physical transmattering in which meaning transformations take place. During my research, as a liminal human being, I analysed my relations with my cells (B lymphocytes and fibroblasts), conceived as post-human liminal lives. I describe them here as post-human because they left the human body—my body—to enter the liminal zone of the laboratory and became the matter of experimentation. My body is my personal matter since I have established an intra-action between what I understand as myself and what I myself feel I am. This epistemological approach to my body as mine convinced me that it was my property for me to use as I wished. As a result, I felt entitled to kill my cells or to donate them and, later on, to use one of my genes to produce food items, which I then shared with people (thus, symbolically sharing my body). The project titled *safe suicide* comprised all transformations of biological matter: from an entire organism into a cell line, from normal cells into immortal cells, from a living cell line into frozen storage or biohazardous trash, and from components of a biological experiment into artistic objects. These changes in physical matter were accompanied by transformations of meaning and language. My liminal project provoked finding new ways of understanding and describing liminal lives (cells). The only way to show the dynamics of these multi-level operations was by keeping an accurate observation journal. Now, it works as an archive that enables me to retrace all the transmatterive steps.

The transfiguring of my personal matter did not end on the level of cell cultivation; it went on to include DNA isolation, the separation of a particular gene, and transferring it from one organism to another. The transmattering stage was provoked by my liminal project titled *The Last Supper*, in which living tissues were transmattered into genetic information (acquiring a new meaning) and then inserted into another living organism. Encapsulated in a plasmid, a gene travelled between my own body through bacterial bodies to the body of a yeast species known as *Saccharomyces cerevisiae*. This series of peculiar intra-actions was also meticulously depicted in my autoethnographic journal.

Translating what was produced in the laboratory into communication understandable in an artistic context required another transmattering of my personal matter and its entanglements. Scientific results had to become art objects or elements of art installations. On the whole, such ventures involve an interesting problem as laboratory protocols for working with biological matter do not fit artistic institutions. Some forms of biomatter cannot leave the laboratory area because biohazard and biopolitics rules ban that (which was the case with *The Last Supper*). Usually, the laboratory environment must be partly remade into an exhibition space. Basically, the transmattered matter cannot travel between contexts alone. It must carry a part of the scientific context along and bring it into the field of art. Another problem with such transfers is the loss of dynamic. The liminal vibration tends to stop on crossing the border of an artistic site, at least in regular exhibitions. Even if a show is repeated several times, it does not mean that transmattering is also repeated. It is rather the matter of multiplied re-presenting. This raises the question whether transmattering ceases when the liminal project is completed. In my own practice, the transmattering process has always been continued in a new project. This continuity has sustained my liminal vibration. Simply, different transmatterive operations have occurred in line with a project's specifications. For its part, the process itself has remained essentially the same. It has always been my personal matter engaged in constant intra-actions. This makes my trajectories unique and unrepeatable by anybody in exactly the same shape. However, every reader can replace my core with their own entangled personal matter, which will start a different transmattering trajectory. They can also interpret and quote the written matter I created while transmattering. Still, it will not be new transmattering since the new—biological and semantic—self will not be reconfigured.

## References


Foucault, Michel. 1977. *Language, Counter-Memory, Practice: Selected Essays and Interviews*. Ithaca, NY: Cornell University Press.

Oppermann, Serpil. 2018. Storied Matter. In *Posthuman Glossary*, ed. Rosi Braidotti and Maria Hlavajova, 411–414. London; Oxford; New York; New Delhi; Sydney: Bloomsbury Academic.

CHAPTER 2

# Transmattering as a Concept in Action

As a liminal being, co-vibrating between several fields, I as a rule tend to catch many ideas that resonate with my fieldwork experiences. My pursuits can thus be characterized as conducting an episodic theoretical narrative without attempting to describe one or another concept thoroughly. However, I do not aim to repeat what has been written many times over; rather, I want to see how these ideas work in new and often-unexpected entanglements.

Harawayan becoming-with is not only about becoming together with other liminal lives; it is also about collective/dispersed agency and being-many. I understood that as a perfect explanation of the entanglements of my personal matter with various forms of matter. This co-existence is a never-ending process of being transformed together. All participants in being-with vibrate simultaneously, exchanging signals and influencing each other's reconfigurations. It means becoming together time and again, without any agent radically changing their location or crossing borders.

To explain how the idea of being-many works in the material world, Haraway has proposed the notion of companion species (2016: 13), inviting her readers to think about particular living beings related to them. I used this theoretical pattern to analyse all my relations established when working on *safe suicide* and *The Last Supper*.

Working in biological laboratories, I could be expected to, instead, employ the theoretical equations developed by Science and Technology



K. Żyniewicz, *The Body as Matter in Art and Science*,
https://doi.org/10.1007/978-3-032-13665-7_2

Studies (STS) scholars who did observations in this cognitive field. In this department, Latour's Actor–Network Theory (2005) was my first discovery before coming across Haraway's writing. The idea of knowledge being produced by a network of actants, that is, actors in action, made perfect sense. The only problem was with the actors' utility, and it was hard to imagine that all actors only existed for the sake of their function. While, from my perspective, the Harawayan companions could suitably be called actants (acting actors), the meaningful difference from Latour's model was that, in Haraway's vision of being, many actors live and coexist just for the fact of being and the process of becoming. Becoming-with is based on sensitivity and responsiveness to others. That is why Haraway makes an important distinction between being responsible and being response-able. All companions—whether human or non-human, and of whatever sizes and scales (2016: 16)—are response-able in many different ways:

> Response-ability is about both absence and presence, killing and nurturing, living and dying—and remembering who lives and who dies and how in the string figures of naturalcultural history. (2016: 27)

I found this passage essential to my research. First, it stressed that even inaction could be seen as a response (the vibration stopped for some reason). Additionally, Haraway counted death as a form of being response-able, which suggested that I was becoming-with my cells even as I was killing them. Being responsive is a general material activity, regardless of what species performs it.

While the response-ability of companion species is a property of collective activity (or inactivity), responsibility seems to be mainly a human task, and it means being responsible for being responsive.

> We are all responsible to and for shaping conditions for multispecies flourishing in the face of terrible histories, and sometimes joyful histories too, but we are not all response-able in the same way. (2016: 29).

Responsibility is evident for humans because of anthropocentric training, which disseminates the idea that the dominant position of human beings in the world obligates them to take responsibility for all other species. Human responsibility is thus related to power and control. Being responsible seems to be easier than being response-able, while only the combination of the two works for the processes of becoming-with and becoming liminal. Humans need to open up to non-human agencies regarding non-human agents as companions, and not as slaves. It is a challenge to the life sciences, which use companion species as research objects. Given this, one of the goals of my liminal practices has been to give scientists an opportunity to be more open, more emotional, and more entangled.

Besides becoming-with and being-with, Haraway's notion of sympoiesis proved pivotal to my work. As Haraway explains,

> Sympoiesis is a simple word; it means 'making-with'. Nothing makes itself; nothing is really autopoietic or self-organizing. (…) Sympoiesis is a word proper to complex, dynamic, responsive, situated, historical systems. It is a word for worlding-with, in company. Sympoiesis enfolds autopoiesis and generatively unfurls and extends it. (2016: 58)

In my understanding, sympoiesis highlights the agency of relational and complex existence, meaning that, rather than a sum of singular agencies of many actors, there is joint agency, a common power resulting from being interconnected. If this approach can be accused of removing individuality and independence, it offers another value in return—the feeling of not being alone and of forming a part of an extensive and well-organized system.

In my imagination, which stems from my evocative practice, sympoiesis evokes the synesthetic experience that arises when we use our senses and share worlding (producing worlds) with other companion species. Being entangled affords us, humans, a chance to extend our senses. We become-with, and we do worlding-with. From my perspective, worlding is an epistemological method. We, all companion matters, make knowledge together because we would be unable to grasp the complexity of the world alone.

While investigating the laboratory world, I had to look at making-with and its participants. Whether all humans doing worlding in the laboratory should be regarded as one species is disputable. It may make sense to treat the practitioners of various disciplines, such as genetics, microbiology, plant biology, and also liminal practices, as different companion species. At the same time, non-human species participate in knowledge production and mostly play the role of model organisms. As Haraway explains:

> A model is a work object; a model is not the same kind of thing as a metaphor or analogy. A model is worked, and it does work. (…) Models in biological research are stabilized systems that can be shared among colleagues

> to investigate questions experimentally and theoretically. Traditionally, biology has had a small set of hard-working living models, each shaped in knots and layers of practice to be apt for some kinds of questions and not others. (2016: 63–4)

If model organisms are strongly objectified and used as tools, one can ask what working with them has in common with worlding-with and becoming-with. As already mentioned, laboratories are rather based on anthropocentric responsibility than on being response-able. Scientists expect model organisms to respond in particular ways; some responses, which are referred to as artefacts, cannot be accepted. The group of model organisms is quite small and includes the bacterium *Escherichia coli*, the fruit fly *Drosophila melanogaster*, the yeasts *Saccharomyces cerevisiae* and *Candida elegans*, the nematode *Caenorhabditis elegans*, the mouse *Mus musculis*, the frog *Xenopus laevis*, the zebrafish *Danio rerio*, the chicken *Gallus gallus*, and the mustard *Arabidopsis thaliana*. This selection was dictated by technical limitations. Working with each of these model organisms requires establishing specific methods. In the light of the Harawayan being-many and being response-able, working with a model organism is a peculiar undertaking. Professional responsibility (for research) makes scientists pre-program the responses of their models. This means that one element of the system tries to force another to meet pre-defined requirements. In this way, we have drawn the full circle back to the anthropocentric attitude.

Biological knowledge is produced using model organisms, but to say that it is made with them as partners on an equal footing would be wide of the mark. Liminal practices try to be better examples of Haraway's vision of the world. However, not every liminal practitioner is aware of her theoretical proposal. Haraway notes that:

> These science art worldings are holobiomes, or holoents, in which scientists, artists, ordinary members of communities, and nonhuman beings become enfolded in each other's projects, in each other's lives; they come to need each other in diverse, passionate, corporeal, meaningful ways. Each is an animating project in deadly times. (2016: 71–2)

These fusions of art, science, and the humanities are focused on being response-able, and they seek to be receptive to all possible responses. They also reflect on sharing responsibility and the belief in the agency and decisiveness of every actor.

Being-many and becoming-with require rethinking and reformulating the notion of objectivity by accepting the body's apparent limitations and acknowledging that other bodies can complement ours. Haraway stresses that the imagined 'we' are embodied entities (1988: 575) and that worlding is fully embodied. Science is not about searching for phenomena that occur in separation from 'us'; it is about vision, which is determined by our sensory system. Further, Haraway proposes embodied objectivity that is processual and individual. Embodied research results in producing situated knowledge. As Haraway insists:

> Feminist objectivity is about limited location and situated knowledge, not about transcendence and splitting of subject and object. It allows us to become answerable for what we learn how to see. (1988: 583)

I perceived situated knowledge as a theoretical and methodological mattering proposal that optimally suited my autoethnographic research. My transmattering was situated because it was based on my personal matter's being-with many other matterial companions. My body and its social and cultural situation were the matter making the transmatterive research possible. I was situated (and vibrated) in the liminal/border zone, registering signals from companions vibrating beyond the disciplinary borders. I was situated in my research as I am present in the text, which is itself matter generated in and through the transmattering process.

Given becoming-with and being many, producing situated knowledge must be founded on split agency (Haraway 1988: 586), which is bound up with being responsive, attentive, and eager to share responsibility, properties I have already evoked. The feminist approach to science calls for rethinking the subject–object duality and in doing so, it presupposes the dispersed agency of multiple subjects. This means that laboratory model organisms should be positioned differently in the laboratory network and recognized as horizontal elements of 'I-with.' In Haraway's words:

> Situated knowledges require that the object of knowledge be pictured as an actor and agent, not a screen or a ground or a resource, never finally as slave to the master that closes off the dialectic in his unique agency and his authorship of 'objective' knowledge. (1988: 592)

As such, situated knowledge promotes seeing the human position in the world differently, or at least to imagine that. It makes it possible to perceive knowledge production not as a strenuous activity performed by the chosen few, but as an embodied experience available to everybody. This kind of research leaves room for surprises and irony (Haraway 1988: 594), and—I would add—for the pleasure of wording and worlding, too.

The Harawayan being-with/becoming-with is in an intriguing conversation with Barad's explanation of the complexity of material life. While Barad's insights into quantum physics and Bohr's rules of mechanics are undoubtedly astute, her ideas about agency are far more relevant to my research. Like Haraway, Barad has sought to offer an anti-dualistic vision of the world and to establish a strong relation between the ontological and epistemological status of things. She has gone even further than Haraway with her being-with and worlding-with and put forward agential realism, which stresses the dynamic and complexity of the material world and its 'worlding':

> I propose 'agential realism' as an epistemological-ontological-ethical framework that provides an understanding of the role of human and nonhuman, material and discursive, and natural and cultural factors is scientific and other social-material practices, thereby moving such considerations beyond the well-worn debates that pit constructivism against realism, agency against structure, and idealism against materialism. Indeed, the new philosophical framework that I propose entails a rethinking of fundamental concepts that support such binary thinking, including the notions of matter, discourse, causality, agency, power, identity, embodiment, objectivity, space, and time. (2007: 26)

Barad's vision brings together elements that have previously been located on the opposite poles. Existing and knowing existence seem to happen simultaneously as one process. Barad's universe, where everything is non-hierarchically interconnected, looks perfect but stirs doubts about its truth. This uncertainty makes this instance of mattering an ideal matter to transmatter, to test empirically. Outlining Haraway's concept of materiality above, I underline the involvement of all actors in constant interactions. For their part, elements of the Baradian nature-cultural network are intra-active, rather than performing interactions. As Barad explains, "intra-actions signify the mutual constitution of entangled agencies" (2007: 33). As already explained, the idea of intra-actions was inspiring for my research since it complemented the notion of 'I-with,' which I constructed based on the Harawayan being-with, becoming-with, and worlding-with. Following the idea of I-with, I would picture interactions as the sum of 'I' and 'you' agencies. Picturing Barad's vision of the material world, I see a system—an incessantly dynamic network in pulsation. This is consistent with my position and self-perception as a liminal being, continually vibrating and calibrating the setting of intra-textual relations.

My liminal practice was profoundly inspired by the organic connection of Barad's onto-epistemology with ethics:

> I explain that ethical concerns are not simply supplemental to the practice of science but an integral part of it. But more than this, I show how values are integral to the nature of knowing and being. Objectivity is simultaneously an epistemological, ontological, and axiological issue, and questions of responsibility and accountability lie at the core of scientific practice. (2007: 37)

This passage highlights the crucial role of science, which I believe, in Barad's vision, is not a practice monopolized by a selected group of people; rather, science is a general way of existing of all the components of the universe. If so, responsibility has an even broader meaning than in Haraway's theory, and it is not an exclusively human domain. Responsibility is imprinted in intra-actions in general. Intra-acting and intra-worlding seem to be a natural, ethical obligation of matter. It is not even about reacting in any given way (as in scientists' expectations of model organisms). It is something ontologically and epistemologically intrinsic to matter. Emphatically, the lack of response does not eliminate an actor from intra-actions. In the account of my fieldwork experiences from the production of *safe suicide* and *The Last Supper*, I will return to intra-activity to explain that the agency of both projects resulted from my intra-actions with a network of human, post-human, and non-human actors.

The performative approach I adopted in my projects also stemmed from Barad's vision of reality, which I construed as an alternative to representationalism. In Baradian onto-epistemology, observing from a distance does not work; knowing is based on a "direct material engagement with the world" (2007: 49) or, as Haraway would put it, on being situated. In my reading, knowing is a nature-cultural mode of existence rather than just an activity specific to science as a genre of cultural production. This point of view promotes respect for all kinds of liminal practices.

Understanding knowledge/science production in this way has a meaningful influence on experimentation. In Barad's view:

> (…) experimenting and theorizing are dynamic practices that play a constitutive role in the production of objects and subjects and matter and meaning. As I will explain, theorizing and experimenting are not about intervening (from outside) but about intra-acting from within, and as part of, the phenomena produced. (2007: 56)

This sisterhood of experimenting and theorizing was pivotal for my idea of transmattering the experimental and theoretical mattering in place. I attempted to intra-act with other existing material elements in a discursive way. The scholarly conceptions described in this chapter also worked as matter to be transmattered through my situated and embodied experience. Experimenting on thoughts and terms was transmattered by translating these operations into experiments done on physical matter (my entangled personal matter). My venture was not about an exact repetition but rather about the initiation of agency through intra-actions of theoretical and physical matters.

Barad claims that knowing does not necessarily entail naming or using verbal and written language; knowing can be embodied. This approach accommodates performative and sensual experiences that are often offered by liminal practices involving art. The Baradian intra-actions are discursive, but discourse is not the sole domain of humans using language or gestures; discursiveness is a feature of matter in general as all its elements intra-act.

An intriguing aspect of Barad's vision of materiality is the positioning of the human body. Because she does not view bodies, in general, as objects with fixed borders and properties, there is no difference between human and non-human bodies in her model. All bodies are "material-discursive phenomena" (2007: 153). I put this approach to the test in *safe suicide* and *The Last Supper* when I first isolated my cells and then the genetic material from them. These operations worked as an extension and dispersal of my body, showing that my I-with could also work as my body-with. The entanglements on the level of my biological and mental self were essential for my transmatterive research.

Becoming-with and knowing-with require a new approach to knowledge production. Latour proposed understanding knowledge production as a matter of concern instead of a matter of fact, which had been the prevalent standpoint in the philosophy of science before the STS movement. Feminist thought has further transformed a matter of concern into a matter of care. Concern and care are related, but "care has stronger affective and ethical connotations." Bellacasa explains that we can readily grasp this by considering "the difference between 'I am concerned' and 'I care'" (2017: 42). Even without any special theoretical explanation, it is easy to realize that caring has a more vital emotional dimension and a more positive meaning than being concerned. Tellingly, being concerned is a passive state, while caring is premised on being active (Puig de la Bellacasa 2017: 42). In Bellacasa's view, caring can be recognized as a transformative ethos of mattering (2017: 67).

In my interpretation, care is an inherent value of becoming-with, worlding-with and being intra-active. In the light of Haraway's and Barad's visions of materiality, relations between agents are based on caring, which is their primary condition and not something emerging for practical reasons. Connections that are founded on caring should be balanced; otherwise, caring can be overwhelming, consuming, and appropriating (Puig de la Bellacasa 2017: 85). Over-caring can result in some agents dominating others. As opposed to this, for humans, proper caring means being simultaneously response-able and responsible. Working in the laboratory taught me how challenging it could be to care without expecting any specific reactions from model organisms and tissues in culturing. The sense of responsibility for the research and the project idea is usually much more potent than being genuinely attentive and sensitive to others' reaction or the lack of it. Like Haraway and Barad, Bellacasa insists that we focus on our 'we-with,' instead of seeing humans as the reference points for all phenomena. I am aware that worlding-with may look like a romantic idea and that it is a challenge both in theory and in practice, but it is worth trying to coexist better.

In Bellacasa's narrative, touch is a notion of being inseparable from care, and it denotes much more than a gesture and a manner of sensing the world:

> Standing here as a metonymic way to access the lived and fleshy character of involved care relations, the haptic holds promise against the primacy of detached vision, a promise of thinking and knowing that is 'in touch' with materiality, touched and touching. (2017: 95)

While touching seems to be a primitive gesture, it is perhaps the best method of becoming-with and thinking-with; it is also a tool of situated knowledge production. Transferred to the laboratory, this approach implies being with all the elements of laboratory equipment, instead of waiting for them to produce disembodied representations. If taken at face value, this idea may sound a bit ridiculous because, obviously, nobody spends time in a laboratory keeping company with PCR machines or incubators. However, being with devices in fact means being aware that they intra-act with human and non-human actors of the laboratory scene. In this context, touch vies with vision and offers a greater engagement in knowing practices. Touching is not limited merely to the hands; touching engages the entire body. As Bellacasa concludes, "I touch, therefore I am" (2017: 99). This life statement significantly impacted my transmattering operations, showing me a real connection with all the human and non-human agents I existed-with and thought-with.

Like Haraway and Barad, Bellacasa points out that knowing-with is not the monopoly of scientists since touching is an intuitive way of being (2017: 105). Referring to Barad's intra-actions, she talks about the central relevance of intra-touching, stressing that: "Reality is a process of intra-active touch" (2017: 114). In her model, touching is the basic intra-activity, and being in touch is about constantly reconfiguring, both materially and as meaning.

I built on Haraway, Barad, and Bellacasa to put together a scheme of my fieldwork experiences and observations. Haraway's narrative taught me to be patient with all my relations as a piece of biological matter and matter understood as meaning. To construct and analyse I-with was very high on the list of my research goals. Barad's vision gave me an idea of how all the elements of I-with work. Understanding intra-actions made me explore agency resulting from being-with, instead of focusing on the separate agencies of all the actors of my I-with. Bellacasa's idea of caring made me sensitive to the intra-active quality of the production of agency. All my autoethnographic notes picture how I practised being-with/becoming-with with utmost attention, sensibility, and even vulnerability.

My resources of theoretical matter to transmatter also contained Bennett with her vision of vibrant matter grouped in assemblages.Bennett helped me to analyse material dynamics. As suggested by her idea of giving "voice to a thing-power" (2010: 2), she sees matter as things, which removes the difference between subject and object. Bennett argues that human power should be understood as a thing-power. She posits that human bodies are composed of many lively and self-organized material things (2010: 9), an idea suggesting that the human body can be viewed as an object. While this standpoint does not differ much from what Barad and Haraway propose, Bennett's spotlighting of the human body was crucial for my research. Her perspective was precious as it somehow managed to oppose both anthropocentrism and mechanistic materialism. Since I was working with pieces of matter derived from my body, the concept of the body as a collective of vibrating, self-organized, and lively elements lays the foundation for *safe suicide* and *The Last Supper*. It helped me to analyse my cells as independent liminal lives and my DNA fragments as lively matter. According to Bennett's vitalism, my body worked as a non-solid, open-ended, and diverse assemblage of lively things (2010: 23–4). This meant that I could replace some of its elements without any trouble. In my imagination, a material assemblage was like an artistic assemblage. The notion of assemblage is also used in art to denote a three-dimensional collage composed of variously sized and differently structured pieces of junk. Despite all their disparities, the artwork components produce a new, balanced value. Something can be removed or added, transforming the assemblage but still retaining its character. In *safe suicide*, I transmattered my body assemblage by transferring one of its components—my cells—to a new assemblage of my artistic project.

Bennett's thoughts about the contribution of food to the vibration and transformations of an assemblage were pivotal to *The Last Supper*. Food is a fascinating form of matter, and Bennett characterizes it as "an assemblage converter" (2010: 42). There is a moment when food enters a body and becomes a new collaborator in it. As digestion happens, the new element is blended with those already collaborating in embodied intra-actions. In Bennett's compelling depiction:

> Eating appears as a series of mutual transformations in which the border between inside and outside becomes blurry: my meal is and is not mine; you both are and are not what you eat. (2010: 49)

I will revisit this idea in the chapters dedicated to the laboratory production and exhibition of *The Last Supper* to reflect on how this approach has influenced the concept of transgenic food and how it has worked with bioethical problems involved in the sharing of such food with other people.

The last thing that I borrowed from Bennet was her vitalist manifesto:

> I believe that this pluriverse is traversed by heterogeneities that are continually doing things. I believe it is wrong to deny vitality to nonhuman bodies,

> forces, and forms, and that a careful course of anthropomorphization can help reveal that vitality, even though it resists full translation and exceeds my comprehensive grasp. I believe that encounters with lively matter can chasten my fantasies of human mastery, highlight the common materiality of all that is, expose a wider distribution of agency, and is reshape the self and its interests. (2010: 122)

I share Bennett's belief in vital matter that is not driven by something non-material (such as the soul) and her conviction that the anthropomorphizing of some assemblage elements can help humans to be attentive to all the self-organized fragments we embody. Anthropomorphizing my cells to a degree was fundamental to the idea of my *safe suicide* project.

An essential aspect of Bennett's vitalist vision is her understanding of affect as something not specific solely to human bodies (2010: xii). Neither an emotion nor an experience, affect is impersonal *conatus* (2010: 2). All material (matterive)—organic and non-organic, natural and cultural—bodies are affective (2010: xii). This idea is indebted to Gilles Deleuze and Félix Guattari, who, themselves inspired by Spinoza, explain that:

> We know nothing about a body until we know what it can do, in other words, what its affects are, how they can or cannot enter into composition with other affects, with the affects of another body...to destroy that body or to be destroyed by it...to exchange actions and passions with it or to join with in composing a more powerful body. (qtd. in Bennett 2010: xii–xiii)

To Bennett, impersonal affect means material vibrancy. It is not something added from the outside as being material is equivalent to being affective (2010: xiii). Understanding agency in terms of affects was crucial to me in analysing the production of my liminal projects, especially at its laboratory stage. It made me more attentive and responsive to every form of materiality I faced, including my own body and the liminal lives derived from my body. The affectiveness of all the pieces of matter offered me an explanation of laboratory and gallery failures. The vision of a joint affective body (Conatus) prompted me to rethink responsibility as shared rather than fully ensuing from me. I-with is affective and responsible.

In my projects, I have relied on the idea of storied matter, primarily developed by Oppermann often in collaboration with Iovino, as a bridge between theoretical and methodological matter. As Oppermann explains:

> 'Storied Matter' is one of the conceptual tools of material ecocriticism, which basically underlines the idea that matter is not only lively, agentic and generative, as it is theorized in the new materialist paradigm (Barad 2007; Bennett 2010; Coole and Frost 2010; Abram 2010), but also densely storied. (2018: 411)

Without thoroughly discussing the concept of material ecocriticism as a whole, I will only draw on one aspect of storied matter that is in close dialogue with my fieldwork experiences. Specifically, 'storied matter' is:

> (…) a material 'mesh' of meanings, properties, and processes, in which human and nonhuman players are interlocked in networks that produce undeniable signifying forces. (Iovino and Oppermann 2014: 1–2)

The notion of a material mesh where meanings, properties, and processes are entangled with each other grounds my idea of multi-level transmattering that unfolds in intra-actions. A transmattering of physical matter automatically brings about a transmattering of meanings and language. The idea does not cancel Latour's Actor–Network Theory but complements it by showing the relational structures from a different angle.

Oppermann stresses that among many material agencies, narrative agency is the one that is crucial to the production of culture (2018: 412). All material entanglements happen as a non-linguistic performance that is imprinted in each participating bit of matter. Constructing a theoretical narrative entails bringing these embodied stories out. Oppermann spells out the implications of this:

> This means that matter's stories emerge through humans, but at the same time humans themselves "emerge through 'material agencies' that leave their traces in lives as well as stories" (Cohen 2015b: 36). (2018: 411)

Humans not only participate in running embodied performances; some of them also attempt to be self-reflective and write material stories down, with the two activities taking place simultaneously. The human body does not stop being an element of storied matter while articulating a part of the story. This may be one reason why writing a narrative seems to be delayed and inadequate to the material story which the body is co-plotting with other forms of matter. The articulation process is always slower than a non-linguistic performance.

The concurrent happening of performing and writing stories implies that the world and the written text (nature and culture) can be perceived as "circulating references" (Iovino and Oppermann 2014: 10). This vision dovetails with my previous discussion of the co-vibrating, liminality, and processualism of I-with. "Matter is eternal life" (Fowles qtd. in Iovino 2014: 21)—I had known this from experience before I came across that statement as a piece of theoretical matter.

The methodological value of the 'storied matter' concept lies in highlighting that research is based on storytelling and on deciphering non-linguistic performances enacted by matter. The art of researching is predicated on discovering and reading these non-articulated stories, on embodied memories.

## References


Barad, Karen. 2007. *Meeting the Universe Halfway: Quantum Physics and the Entanglement of Matter and Meaning*. Durham and London: Duke University Press.

Bennett, Jane. 2010. *Vibrant Matter: A Political Ecology of Things*. Durham and London: Duke University Press.

Haraway, Donna. 1988. Situated Knowledges: The Science Question in Feminism and the Privilege of Partial Perspective. *Feminist Studies* 14 (3): 575–599.

Haraway, Donna. 2016. *Staying with the Trouble: Making Kin in the Chthulucene*. Durham and London: Duke University Press.

Iovino, Serenella. 2014. *Ecocriticism and Italy: Ecology, Resistance, and Liberation*. London and New York: Bloomsbury Academic.

Iovino, Serenella, and Serpil Oppermann. 2014. *Material Ecocriticism*. Bloomington: Indiana University Press.

Latour, Bruno. 2005. *Reassembling the Social: An Introduction to Actor-Network-Theory*. New York: Oxford University Press.

Oppermann, Serpil. 2018. Storied Matter. In *Posthuman Glossary*, ed. Rosi Braidotti and Maria Hlavajova, 411–414. London; Oxford; New York; New Delhi; Sydney: Bloomsbury Academic.

Puig de la Bellacasa, María. 2017. *Matters of Care: Speculative Ethics in More Than Human Worlds*. Minneapolis and London: University of Minnesota Press.

CHAPTER 3

# The Liminal Zone of Laboratory

The biological laboratory was part and parcel of my liminal practice as a territory where liminal lives existed and a space where many mattering operations occurred. I needed guidance to construct my autoethnography recording the production of liminal projects in biological laboratories. New Materialist ideas prompted me to consider the entirety of material reality. Since laboratories were crucial for my research, I realized that I needed to study them as specific, peculiarly constructed segments of materiality. The terms 'constructed' and 'created' are essential in this context. For some reason, laboratories differ from the general production of culture, and they are distinctively framed. From my perspective, their operations are reminiscent of theatres with their abundant representations.

I needed some protocols to help me to navigate the laboratory stage. It was essential for me to reflect on the position I took there. Did I play a main role, or did I only work as a prompter's voice, coming from the liminal zone? Having decided to carry out an autoethnographic study and try to identify a foreignness in my I-with, I had to name and depict its transformation into you-with.

Some laboratory ethnographies conducted within the Science and Technology Studies (STS) paradigm have drawn up models of the laboratory construct/mise-en-scène. Latour's Actor–Network Theory (ANT), the most thorough account of which is to be found in his book *Reassembling the Social* (2005), is not at odds with the theoretical approaches to materiality discussed above, and the notion of the actant/actor has already



K. Żyniewicz, *The Body as Matter in Art and Science*,
https://doi.org/10.1007/978-3-032-13665-7_3

been evoked in conjunction with becoming-with, worlding-with, intra-actions with care, and a material mesh. I owe to Latour a helpful name for the multiple agencies co-creating I-with.

In my research, the Latourian concept of network helped me to imagine the laboratory fieldwork graphically and to draw up a map of laboratory reality, following which I could remain sensitive to all possible agents. It was also a map that I tried to transmatter by incorporating art and artistic agents.

Latour stresses that laboratory knowledge production has a relational character. Several actors exist in the network or, to put it differently, are unable to live without the web. This view is consistent with the Baradian idea of intra-actions. They are actor-networks, stimulating each other's activity (Latour 2005: 46). Latour defines an actor in action as an actant. His theory helped me to solve a puzzle that I encountered when trying to work with feminist New Materialism. I was unsure how to handle non-living agencies, such as laboratory devices, measurement machines, and all other laboratory appliances. Haraway's writings rather concern human and non-human living actors (companion species). While her insights were central to my projects, it seemed misguided to treat all the non-living items as companion species. The idea of intra-actions was easier to apply to non-living agencies. Science is an agency resulting from the intra-acting of multiple living and non-living actors. For its part, ANT helpfully takes into account living and non-living actants, revealing the complex structure of social processes. I took Latour's emphasis on the importance of the network as a methodological guideline to identify all the threads of the network. I firmly relied on the Latourian theory in founding my laboratory research on the investigation of the network and on analysing my position in this relational structure. I learned to see the laboratory as a collective of actors that played a realistic 'theatre piece' about the complexity of the material world. As Latour has pointed out, science works on *in vivo* complexity in *in vitro* cultures (1993: 61). I found it essential that Latour disconnected science from nature, which is the subject of scientific operations, and asserted that culture was an artefact: "the very notion of culture is an artifact created by bracketing Nature off" (1993: 104), meaning that humans could never know nature itself. Nature is only our cultural production, based on material stories written by humans, as Oppermann and Iovino would put it. When working in biological laboratories, I was always convinced that creativity played a crucial role there, which made me feel that I could intra-act effectively in this environment. However, I was there as a visiting actress working for a different 'theatre piece.' I could try and join the cast, but I felt awkward wearing a different costume and having learned a different script. It took me some time to learn how to play with others and, to a certain extent, I was able to mimic their movements and language. However, as soon as I played my episode, I got back to my liminal theatre.

Latour explains that a network is an instrument for investigating something, instead of being something to be investigated (2010: 188). I understood this explication as communicating that he had built his own theatre based on laboratory observations. That gave me the comforting belief that I had the right to develop a theatre of my own in my research. I used the experience of being a visiting actress to direct my pieces and was responsible for my version of the mise-en-scène.

I also took for myself Latour's insights into constructing and understanding texts, specifically his idea of the writer reconstructing the network or creating a new one based on actants' representations. It is the new theatre. Understanding writing as a reconstruction effort was very encouraging as it made room for trials and errors. Even being responsible does not guarantee that the writer will intra-act with all readers in one specific way. Seeing my writing as a new theatre of representations where academic actors were hired helped me to be more sensitive to them.

Latour's vision of laboratories attributes a prominent role to representations. He insists that science re-presents reality by means of various scientific methods (1999: 30). The knowledge produced is entirely constructed and constituted by the laboratory network. When doing my laboratory fieldwork, I came up with three critical questions: Do I produce representations of representations (second-hand representations) while implementing liminal projects underpinned by scientific methods? Does the art context breed a different representational language? What is the role of representation in the transmattering processes? The answer to the first question was affirmative. To build my theatre, I had to superimpose artistic representations on the scientific ones. The answer to the second question was likewise affirmative. The art context did engender a non-textual language to use. In general, a robust artistic operation is an aestheticization that often transmatters discarded elements of the laboratory set-up into something appreciated. Finally, the answer to the third question was that the role of representations in transmattering processes was fundamental. My personal matter was transmattered into laboratory representations immediately after the isolation of my cells. Subsequently, these representations were transmattered many times in the laboratory and art spaces.

Besides all these concepts, I took from Latour the notion of 'laboratorization', on which he dwells, among others, in his paper '*Give Me a Laboratory and I will Raise the World*' (1983). Laboratorization denotes an expansion of laboratory conditions outside the laboratory as such. In this move, the laboratory turns into an open-air theatre or, more precisely perhaps, a performance engaging an audience. Latour's example is Pasteur's anthrax research, in which Pasteur included very important actants, or rather brought into relief their presence, in society: microbes (1983: 169). This gesture removed the 'inside'/'outside' divisions pertaining to both the laboratory and the body. As Latour observes, Pasteur had to 'buy' public attention for his research by sparking social interest. Today, his activity would be regarded as science popularization. Pasteur had to do a 'staging'—a performative presentation (1983: 151)—which took some other skills than those necessary for laboratory work. Given this, his venture resembled liminal practices, which are usually conflated with science popularizing. However, their goal is to share the knowledge and foster critical thinking and social dialogue beyond disciplines and institutions. Laboratorization has undermined the belief that laboratories are selected segments of material reality, where only particular actors can intra-act. The laboratory has spilled over the whole of reality. I applied the idea of laboratorization, mostly exploring its extension: art galleries and humanities departments.

Pasteur's laboratory research and the microscopic technology made invisible actors visible, stressing their importance for the research itself and for much broader processes. Pasteur proposed replacing the notion of society with one of a collective (Latour 2013: 242). That marked a breakthrough from the perspective of today's humanities and art. Even though microbes were still primarily seen in terms of their function, they started to be recognized as intra-acting with humans. Liminal practices, including mine, seek to show non-human agents for themselves and to encourage humans to respect them, consciously care about them, and learn how to become liminal with them. The artistic aspect of liminal practices promotes developing an emotional approach to non-human laboratory actors. They are no longer mere elements of the setup, and they simply play.

Biologists working with microbes and applying Pasteur's laboratory research approach faced the challenge of carrying laboratory conditions over into general reality, which consequently came to resemble one extensive laboratory (Latour 2009: 189). Liminal practitioners working on biologically-grounded projects do something similar, but with regard to the laboratorization of the art world. The biologically based field of art is part of a general artistic environment, but its relation to laboratory conditions has not yet been precisely defined. Exhibitions of liminal projects often require constructing a typical laboratory infrastructure. As a result of this laboratorizing activity, non-human laboratory agents are not only our partners in daily routines but also culture (art) creators.

Latour distinguishes fetishes, facts, and factishes as very peculiar actors in the theatre of representations that helped me build my liminal theatre programmes. Considering the etymology of 'fetish' in French, Latour explains that it means a "fairy object" and points out that it is a "talk maker" (2010: 3–4). In my understanding, a fetish is a physical or imagined reference point for plotting stories around it. Religions are perfect examples of such narratives. There are people that believe in fetishes and those that do not. The latter—anti-fetishists—perceive a fetish as an overhead projector and deem it a human thing covered in a non-human camouflage (2010: 8).

In his assessment of the Moderns, Latour emphasizes that they tend to disconnect fetishes and facts, an inclination related to the division between knowledge and illusion, and "between practical life which does not make this distinction, and theoretical life which maintains it" (2010: 11–2). Anti-fetishists are also given a different moniker in Latour's theory: they are called critical thinkers or laboratory knowledge constructors. They believe in facts, and, given the etymology of the word 'fact,' facts denote something that is fabricated and at the same time not fabricated (2010: 18). Facts are fetishes of science. After being fabricated, they become independent of the knowing/fabricating subject (2010: 18).

From Latour's point of view, cherishing the value of facts while deprecating the value of fetishes does not make sense. They are both fabricated, and they both take control of their constructors. The word 'fact' points at something external, while 'fetish' is associated with an internal belief of a person (2010: 21). People tend to not trust their personal visions and experiences, and so they feel an urge to construct an external 'truth.' The laboratory works as a place dedicated to the production of the truth.

Taking into account the common agency of fetishes and facts, Latour proposes a third notion: factish. A 'factish' is something that is fabricated and becomes independent as a thing in which people believe and without which they cannot live. In my reading, the conclusion of Latour's reasoning is that everything that humans know about their life and its circumstances is created by them. There is nothing that people know that is not fabricated. In this light, science is a story plotted around facts, and art is a story built on fetishes.

Plotting stories around factishes is exactly what liminal practices try to do. It is building evocative, personal, and emotional stories on scientific facts, manufactured in scientific laboratories. The issue is that art and the humanities are suspicious of the power of facts, and scientists do not believe in the agency of fetishes. Dismissed by Latour, the modern tendency to rely on divisions is still very much present in the institutional production of culture. The life and hard sciences continue to be thought of as important because they purportedly reveal the truth while other research fields are viewed as creating 'fables.' The emergence of transdisciplinary and interdisciplinary practices is a harbinger of possible changes in the system. All 'in-between' activities substantiate Latour's vision of factishes and lend it credibility.

Besides Latour, I am indebted to Knorr-Cetina and her definition of epistemic cultures (1999), which holds that epistemic cultures are those "that create and warrant knowledge, and the premier knowledge institution throughout the world is, still, science" (1999: 1). Rationality is an essential aspect of examining the production of knowledge. Knorr-Cetina observes that: "Philosophers' theories of science are generally theories of scientific rationality" (1991: 105). The problem, as Knorr-Cetina sees it, lies in ignoring the relevance of the context, as philosophers of science fail to consider personal, social, and broader cultural factors alike (1991: 106). Knorr-Cetina believes that this is a serious misconception, to prove which she has collected evidence corroborating the centrality of context and its construction in epistemic processes: "Context or, in a broad sense, 'culture' is inside the epistemic" (1991: 107). Searching for patterns and strategies used in knowledge production, Knorr-Cetina did observations in two types of scientific laboratories: particle physics and molecular biology ones. This helped her to show meaningful differences between these two sites of epistemic production. In physics, she has identified two main strategies: a framing strategy and an exchange strategy. The former is mobilized "to cure reflexivity and similar problems through considering comparable data and objects which serve as a frame of reference in dealing with the problem" (Knorr-Cetina 1991: 111). In the exchange strategy, scientific reasons are objects sold for money and other non-scientific goods. This influences the atmosphere of knowledge production, making it highly competitive.

For its part, molecular biology relies on different strategies. First of all, reasoning and searching for solutions are delegated to individual scientists: "individuals are expected to synthesize from their own earlier experience the prognostic knowledge they need to repair circularities and contingencies" (Knorr-Cetina 1991: 115). Individual engagement lies at the core of what Knorr-Cetina dubs a biographical procedure which is closely intertwined with two other basic biological laboratory procedures: a chaotic procedure and a systematic procedure. These procedures are located on the two opposite ends of the scale, ranging from well-planned and ordered experiments to messy multitasking without proper coordination.

Another meaningful difference between the epistemic systems of physics and biology is dealing or not dealing with physical matter. Knorr-Cetina's observations imply that in particle physics, scientists deal with signs or signatures of phenomena rather than with the phenomena themselves (1991: 117). Their experimentation is based on "a network of anticipation, simulation, and recalculation" (1991: 118). The molecular biology laboratory cuts quite an opposite figure. It appears to be based on contact with the empirical world since "(quasi-)natural objects are sought out and encountered on a day-to-day basis" (1991: 119). The laboratory work is founded on taking care of, examining, and experimenting with matter. As Knorr-Cetina puts it:

> scientists and 'things' in the molecular genetics laboratory are joined together in interactional/behavioral circuits, in systems of conjoint reaction and reciprocity in which neither behaves independently of nor completely dominates the other. (1991: 119)

Recognizing meaningful differences between epistemic systems (in physics and in biology), Knorr-Cetina underscores that that each of them is "a richly textured internal environment and culture" (1991: 120). Epistemic cultures are internally structured and organized; they definitely resemble theatres producing rationality plays. Knorr-Cetina's insights made me aware that I would join quite hermetic setups which appointed every actor a particular role to play and were organized as a meaningful hierarchy, rather than being horizontally structured. Knorr-Cetina's portrayal of laboratories also reminded me of Plato's cave, where humans could not see physical objects but were part of a shadow theatre. This approach encouraged me to take on laboratory actors to play in my artistic script. They were transmattered by my artistic context. These observations were highly important for my autoethnographic research in biological laboratories. Knorr-Cetina's depiction of the molecular biology laboratory corresponds to Latour's Actor–Network Theory. They both perceive that which is going on in the laboratory as a system of interactions involving many different human and non-human actors.

However, the visions of the laboratory offered by Latour and Knorr-Cetina are different, although they both stress the complexity of agency. Knorr-Cetina's laboratory is not a Latourian network; it is a mechanism. In conceptualizing the position of the scientist, Latour regards the researcher as one of many actants, and Knorr-Cetina considers scientists to be "methods of inquiry; they are part of a field's research strategy and a technical device in the production of knowledge" (1999: 29). This assessment did not make me respect scientists less, but I felt that I also could find a place for myself in the structure of the laboratory, at least for some time.

In Knorr-Cetina's constructivist thought, the laboratory and all laboratory organisms and liminal lives are seen as "molecular machines" (1999: 148). These small machines consist of many replaceable and removable components. This notion was confirmed in my fieldwork observations when a geneticist tried to explain to me that cells were only screws. This is even more patently visible in synthetic biology, which is based on manipulable biobricks capable of being arranged in many different ways. The mechanical structure of laboratory life as portrayed by Knorr-Cetina has its meaningful levels. First, as she points out, molecular biology views life as self-reproducing biological machines. Subsequently, these machines are reproduced in the laboratory, optimizing their work and adjusting them to the research plan, as a result of which they become 'purified' versions. Finally, these optimized mechanisms work as models/representations of the original mechanisms (1999: 157). Given that the laboratory works with traces of natural objects, biology is unavoidably a science focused on life but performing without nature as such. My goal was not to find out whether Knorr-Cetina was correct or not; rather, I wanted to investigate how her vision of epistemic cultures worked for my liminal practice. I sought to see what would happen to this mechanical perspective when laboratory conditions were moved to the exhibition context. Was it going to be the same theatre or maybe the same scenography transferred to a new one?

Last but not least, I incorporated Knorr-Cetina's analysis of the economy of care into my research. This meaningfully tied in with Bellacasa's model of care. Knorr-Cetina observes that:

> the care extended to cells is motivated not by moral consideration but by an economy of time and resources. The concern for their well-being is a concern for the work and materials wasted if things go wrong. (1999:: 145)

This insight was illuminating, especially when I 'killed' my cells while working on *safe suicide*. Every element of the laboratory setup plays a very particular role. They do not exist 'just because,' though the New Materialists would argue just that; they exist because they represent something else.

I collected many experimental results that were not useful from the point of view of my liminal project and thus of no interest to me, if not simply unacceptable. They were as a rule visually intriguing, even decorative, but could not be a source of valuable data. I had to find a category to capture their position in my laboratory work. Michael Lynch's research and his definition of laboratory artefacts were of great help to me in this respect. As Lynch puts it:

> Artifacts are described as particular 'intrusions' or 'distortion' in the observability of the 'natural' features of the world. They are derived from the instrumental conditions of their perception. The danger that artifacts are said to present for research is that their presence can go undiscovered and be taken as evidentiary features of purportedly natural phenomena. (1985: 81)

In my view, laboratory artefacts can be regarded as the technological abject. I am not interested here in the psychoanalytical aspect of the abject; what is relevant to me is its expulsion from the epistemic machine. Artefacts are outside the order, and they cross the limits of cognitive expectations. Abjectual data result from glitches in the system. They can have the character of regular trash or junk, which can be helpful at some moments. Recognizing artefacts is a matter of a scientist's individual interpretation. Artefacts can result from mistakes in specimen preparation (Lynch 1985: 89) or from device malfunctions. Some of them are obvious because of poor visibility or bad quality in general, but some can be tricky to identify as artefacts. It is an issue to discuss for the laboratory team whether a given result is good enough to use in the research, or whether it should be eliminated (Lynch 1985: 96). In the light of ANT, artefacts are regular actants in the laboratory network. Even though they are not wanted, they trigger certain actions and force human agents to make certain decisions. In terms of my metaphor of the theatre of representations, artefacts are the villains of the plot. Returning to Bennett's vision of impersonal affects, I would explain artefacts as traces of the effectiveness of material actors. The false scientific results are evidence that humans are part of the Conatus, and that they unable to control fully the attentiveness of the other parts.

Biologists often think that the visual appeal of artefacts makes them more artistic than regular results. It was the matter of my decision to treat artefacts in the same way as scientists did. I collected artefacts as a separate category of visual fieldwork materials that were not included in the gallery presentations of my liminal projects. It was in fact much easier to obtain an artefact than a regular, valuable result, and that was evidence that laboratory work required many trials and considerable resistance to failure. Examining the same artefact with more than one scientist also helped me to realize how subjective the interpretation of results could be.

Throughout this text, I call myself a liminal practitioner and refer to my practice as a liminal practice (and to my projects as liminal projects). When employing the notion of liminality, I rely on the tradition of anthropological and social research focused on the rites of passage, especially on Arnold van Gennep's studies of the tripartite structure of the rite, as continued and developed by Victor Turner.

Bjørn Thomassen, in his book *Liminality and the Modern: Living Through the In-Between*, carefully examines transformations of liminality from the time described by van Gennep to our post-modern age. His detailed work helped me to understand and explain my liminality and its meaning for the transmattering I performed in laboratories and outside. Thomassen's general definition of liminality holds that:

> Liminality refers to moments or periods of transition during which the normal limits to thought, self-understanding and behaviour are relaxed, opening the way to novelty and imagination, construction and destruction. (2018: 1)

Further, Thomassen points out that the relationship between structure and agency is unclear in the liminal state, which is an uncertain condition without any determined status. His account of liminality comprises its individual and social aspects, both of which were crucial for my transmatterive research. An essential question that I had to address in my study was: Does transmattering only happen in liminal conditions, or does it concern pre-liminal and post-liminal stages as well? A legitimate question to ask at this point is also how matter in the pre-/post-liminal zone can possibly intra-act with the liminal being if pre- and post-liminal conditions are characterized as static. The idea of intra-vibrating comes in handy as an answer. The vibration of the liminal being initiates the vibration of matter in each bordering zone. The agency of the liminal being results from co-vibrating. This makes all the forms and elements of matter constantly move and recalibrate without crossing the borders. Emphatically, being liminal means being located nowhere; the border/liminal zone is a space equal in status to all the neighbouring spaces. The objective of my research was to prove that the liminal zone was not a black spot or just an empty field to pass on the way from one border to another. The liminal position supplied transmatterive power for the reconfiguration of I-with through intra-actions with foreign matters.

In his first book, *Les Rites de Passage* (1909), van Gennep proposed the tripartite structure of ritual, identifying three sub-categories of rituals: rites of separation, ceremonies of transition, and rites of incorporation, which correspond to pre-liminal, liminal, and post-liminal stages, respectively (1960: 11). During the first stage, an individual or a group are situated in a well-known environment/condition; in the second stage, they undergo a transition, which is also conceptualized as the phase of in-betweenness. A passing ritual is perfectly exemplified by the traditional practices marking boys' entry into manhood or warriorhood, in which youngsters have to leave their safe home environment and parental care in order to go through the liminal stage of trials, the successful completion of which stands as evidence of their manliness. The mysterious liminal phase of tests is full of hope and curiosity, but also of uncertainty and fear. The last—post-liminal—stage is like a return to calm and safety because an individual or a group enter a new environment/condition, which is structured, defined, static, and boring. Van Gennep's identification of this ritual structure was crucial to the development of anthropology and the social sciences. The theory is more than one hundred years old, but over the last century it has invited exciting reinterpretations and adjustments to new historical circumstances. I used van Gennep's approach to transmatter it and coin my own definition of liminality and liminal location.

To transmatter the basic definition of liminality, I needed to retrace all stages of its mattering. As already mentioned, Turner upgraded van Gennep's findings by moving the rite analysis to modern conditions. The questions that he sought to answer were: "What happened to liminality within a horizon of the modern? Did it get lost? Did it survive? Did it mutate into something else? If so, then what?" (Thomassen 2018: 82). Turner provided answers in his article 'Liminal to Liminoid, in Play, Flow and Ritual: An Essay in Comparative Symbology' (1974), suggesting that ritualistic experiences in consumerist societies had been replaced by liminoid experiences, which primarily took place in "art and leisure activities" (Thomassen 2018: 82). Turner acknowledged that liminoid events had their source in liminality but insisted that the two conditions differed significantly. Liminality was characteristic of tribal communities, while liminoidity referred to modern capitalistic societies. Traditional rites were intimately associated with the cycles of nature and thus regularly repetitive; for their part, modern rites were generated and not necessarily reiterated. The lack of exact repetition was crucial for my liminal practice. When considered as liminoid situations, the occurrence of exhibitions and art performances is not related to the natural annual rhythm. It is only very rarely that an artistic project is put on display every year. Additionally, every artistic event is different, and an artistic project is situated in a different space and in the company of different artworks on every occasion. Th effects of such liminoid experiences are also unique, unlike in the rite of becoming a man, where the presupposed result was, exactly, becoming a man.

As Thomassen observes, some departures of Turner's proposition from the original scheme of ritual offered by van Gennep were rather disputable. First, the liminoid experiences as conceptualized by Turner do not include an aspect of danger and destruction, which were fundamental to van Gennep's notion of liminal processes. What Turner chose to focus on was the performative and aesthetic character of liminoid events, which has made his research mostly instrumental in the field of art and performance, but less so in the social sciences and anthropology. Second, Turner's take on *communitas*, a social formation helping individuals pass the ritual, is overly romantic in insisting that communitas is based on love and shared understanding whereas, in the basic rite model, communitas could in fact be founded on violent and negative feelings as well. Finally, liminoid experiences are optional, playful, and entertaining, which means that they cannot be seen as a transformative transition entailing border crossing and discomfort, experiences intrinsic to ritual (Thomassen 2018: 83–4). In my view, liminoid events are rather representations of transitions and of liminal experiences. All these doubts point at what can be perceived as the loss of ritual power. Contemporary societies no longer experience a real catharsis. Liminoid transformations are safe. Nevertheless, even though partly agreeing that liminoid events do not (or at least should not) occasion actual risks to one's life, I believe that, if good, they can be profoundly touching and transformative.

When the transformation of liminality described by Thomassen arrives at the threshold of the postmodern age, liminoid experiences are replaced by limivoid events. Such moments come about when people consciously search for intense experiences, for instance, in extreme sports (such as bungee jumping). The person experiencing a limivoid moment is usually surrounded by friends, family, and/or other people waiting to 'jump.' Bungee jumping spots are located in designated and socially acknowledged places, which are often associated with entertainments and festivities. Additionally, sharing the photo and video documentation of the jump with a broad online audience on social media is an expected part of the venture. Some bungee jumps mark celebrations of transitions in life, such as getting married and graduation (which brings to mind the rite of becoming a man). At the same, to see a bungee jump itself as a formative and transformative process would certainly be an exaggeration. First of all, the experience is compressed to a few minutes. Even if it involves crossing the border of fear, jumping into the void, and risking one's life, one is certain that the jump is supported enough to ensure the return to the safe condition and the restoration of the average adrenaline level. It is coming close to death but without crossing the line. Postmodern societies are not patient enough to undergo lengthy ritual processes; rather, they look for fast 'kicks' and quick boundary experiences. Thomassen also believes that attitudes to liminality are split in contemporary societies. On the one hand, people pursue borderline experiences, nearly to the point of living on the edge; on the other hand, these quick adrenaline 'shots' do not suffice to transform their lives in a meaningful way. People strive to experience a profound transformation, but they are simultaneously terrified of it (2018: 189).

From my perspective, the 'void' element of limivoid experiences is meaningful and intriguing. It operates as a territory beyond the most distant of possible borders, where nothing exists. I am even inclined to see limivoid moments as trials of dematerialization and the annihilation of physical matter together with its meaning, where all intra-actions are broken. It can be a symptom of the over-mattering of the contemporary world, with its superfluity of things and meanings. Jumping into the void offers a moment of de-mattering and freedom from matter, but it is essential not to cross the border and to continue to vibrate in the liminal zone. An opposite interpretation is possible as well: as reality is a void that lacks meaning and feelings, individuals need to recalibrate their liminal vibration into a shock.

The above considerations on—and insights into—liminality as such helped me to understand my own liminality, which was my general cognitive attitude in my research. I defined myself as a liminal being, and I participated in a broader process of becoming liminal along with many other human and non-human liminal beings. Importantly, becoming liminal does not in general exclude having liminoid and limivoid experiences, which are interrelated types of liminality that have been spawned in the historical process of civilizational changes.

My journey of becoming liminal has been crucial for my knowledge production and my position in this text. Many years ago, I passed the ritual of becoming an artist by studying at the Academy of Fine Arts and exhibiting at artistic institutions. That was a socially regulated and approved transition from being a young person without any precisely defined function to being a member of a particular profession. Subsequently, I decided to engage with the life sciences and research, without leaving the artistic field. However, I have not fully immersed myself in these new environments, as I have retained some of my previous status, which has blurred my position. The main aspect that distinguishes me from many other liminal lives is that I am aware of my liminality. The liminal zone is my 'home zone', where I am either 'not only' or 'not yet'—not only/not yet an artist, a researcher, and a laboratory practitioner. I step back into art when needed and enter research and science (mostly biology) when necessary. Nevertheless, I always return to the liminal zone, intra-acting at a crossroads. Strategically speaking, this is the best territory, without any definitive declarations and with a backpack full of useful tools wherever I go. Becoming liminal or, more to the point, being aware of my liminality makes me flexible and accessible. It helps me to carry out experimental transmattering operations. Maintaining the 'not-only' and 'not-yet' status challenges all agents involved in my procedures that vibrate with me. They need to come up with new criteria and new definitions to perceive their liminal potential.

Becoming liminal is 'contagious' in that everybody that deals with liminal practitioners to some extent discovers their own liminality. This represents contamination as comprehended by Anna Lowenhaupt Tsing:

> We are contaminated by our encounters; they change who we are as we make way for others. As contamination changes world-making projects, mutual worlds—and new directions—may emerge. Everyone carries a history of contamination; purity is not an option. (2015: 28)

All human and non-human liminal beings become liminal together. Transmattering is always collective and relational. If Tsing avers that self-sustained creatures with individual attitudes are resistant to transformation (2015: 29), my view is that their transmatterive becoming-liminal happens at a slower pace. Since being contaminated by each other enforces an acceptance of diversity (Tsing 2015: 31), the concept of contamination appears to dovetail with the notion of intra-actions.

I invariably believe that the border/liminal zone is constituted and stable in its transmatterive power. Notably, besides revelling in the flexibility and freedom of experimenting and making mistakes, the liminal zone is also encumbered with costs, such as uncertainty and precarity. Institutionally speaking, existing in a liminal area still as a rule means being nowhere, being 'not only' and 'not yet' in terms of the social taxonomies in place; consequently, becoming liminal means being nobody. Liminal human lives stubbornly defy fixed divisions and limits. Becoming liminal is not accounted for in institutional systems, and there is no university department or discipline officially dedicated to that; neither is there a corresponding category in fellowship and grant applications. Sooner or later, liminal practitioners have to inscribe themselves into one disciplinary structure or another which only partly overlaps with their exploration terrain. This means being artificially situated across one of the borders.

In my liminality I am usually too scientific for artists and art audiences and too artistic for academics and biologists. People from the world of art tend to note the creative aspect of my practice; meanwhile, scholars and scientists are mainly focused on the research facet of my pursuits. Everybody selects an aspect that is close to their practice, which seems safer and more accessible. This can be regarded as a form of hygiene, as protection against contamination by uncontrollable liminality and diversity, and as the persistent social denial of inevitable changes.

Introducing myself as an artist or a researcher (in an accurately defined area) is much easier than explaining that I am 'not only' and 'not yet.' Being 'not only' and 'not yet' can be emotionally destructive because of uncertainty and the lack of understanding. This experience is similar to the danger experienced by the participants of primitive rituals, which involved combat or exposure to extreme conditions. However, the members of old tribes were assured that having gone through the liminal stage, they would enter a safe and structured site of post-liminality. I have decided to give up on safety and certainty for the sake of excitement and cognitive pleasure, and I want to contaminate others by accepting vibrating frictions.

Turner described communitas as a natural way of facilitating rites of passage. It seems evident that being in a given situation with other people is usually easier than being in it alone. This observation reverberates in Tsing's statement that "[w]ithout collaboration, we all die" (2015: 29). As a liminal practitioner, I have established a communitas with many other liminal practitioners worldwide. My communitas is the art&science liminal community of people whose lives and work are usually disconnected from disciplinary environments. If every discipline has a professional community of its own, a liminal communitas is based not only on professional collaboration and knowledge exchange but also on friendship and emotional support. Liminal conferences afford opportunities to share feelings and opinions along with findings; such events bring together people from academia, independent artists and researchers, and passionate activists. Not-only/not-yet practitioners, whose notorious predicament is to explain their exact position on the profession's map, can finally relax when talking to people who grapple with similar issues.

Turner defined modern liminality as liminoid. He transposed ritualistic tribal experiences onto the experience of literature and performative art, where liminoid events happen in safe, 'artificial' circumstances. Given that I continue to co-vibrate with the artistic context, one may argue that I am a liminoid being rather than a liminal one. My take is that I am indeed a liminal being that creates liminoid situations in exhibitions, performative events, and workshops. They are related to the general liminality that I represent, and they can be viewed as a manner in which conscious liminality contaminates others. When creating liminoid situations, I take part in them and carry out participant observations. These observations transmatter the events through the lens of my interpretation. The subsequent process of writing transmatters these situations all over again.

My transmatterive existence comprises all forms of liminality listed by Thomassen, including limivoidity. I believe that the development of liminality is largely powered by the shortening of time. Since tribal rituals were related to the seasonal dynamics of nature, it took time to pass a ritual. Liminoid experience is extrapolated onto art and thus seems shorter and repeatable only to a certain extent. Finally, limivoid experience is just a brief moment, a fast-food rite. Remarkably, liminal acts have also become more and more exclusive. While entire communities experienced their tribal rituals, liminoid events bound up with 'high culture' are only accessible to selected members of the public, and limivoid rites are undertaken by even fewer people. Not everybody is a bungee jumper or a fan of extreme sports. However, we are all steeped in capitalistic reality, which profoundly impacts us by promoting fast-food, fast-science, and fast-rites. Being a liminal practitioner who creates liminoid events is a form of limivoid experience because every exhibition and performance is a risk in and of itself, which I believe is comparable to bungee jumping. Every project presentation is like plunging into a void, a dive that elevates one's adrenaline level. As an average member of post-modern society—a society of control, as it is—I feel like that time and again. Becoming liminal requires adjusting to the rite addiction.

Having briefly described the concept of liminality and its historical transformations, I must illumine its biological aspect to properly explain my own liminality, and picture laboratory as its zone.

Interestingly, van Gennep's classic research was pervaded by an interest in biology sparked by his experience in preparing microscope slides. He was mainly fascinated with the methodology of the life sciences, including straightforward observations, systematic data collection, and examination of living facts, which he wanted to apply to the social sciences (Thomassen 2018: 44–5). Nevertheless, the natural mode of observing social processes did not lead van Gennep to incorporate non-human living beings into his study. Such an inclusion is essential in contemporary research, as exemplified by Squier's *Liminal Lives: Imagining the Human at the Frontiers of Biomedicine*. Squier focuses on becoming liminal and emphasizes that ours is the time of multidimensional transformations of embodiment. The body (not only human) finds itself in the process of transmattering from the *in vivo* to the *in vitro* conditions and the other way around. At the same time a massive transformation of the meanings of the body and body discourse is taking place. Squier's vision inspired me to analyse this multilevel transmattering.

In Squier's understanding, we, humans, share liminality with liminal lives or, more precisely, they share it with us. In the introduction to her study, Squier explains that:

> The liminal lives that are the subject of this book exist in the in-between or marginal zone. Like the Nightlight 'adoptable embryos,' neither discarded bioproducts nor valued human beings, they are participants in a rite of passage, between everyday life and a higher or different level of existence. (…) Like the embryos, the embryonic stem cells are also marginal (either temporally or taxonomically) to the human being. (2004: 4)

Besides the living agents mentioned in this passage, Squier also considers normal stem cells and organ transplantation, which all come from human bodies and are transferred to other bodies. This liminal stage takes place in biological and medical laboratories. Time and place are relevant in this venture since ritual experiences have their seminal temporal and spatial dimensions. Van Gennep described the ritual as a territorial passage, distinctively involving the crossing of borders, border zones, thresholds and portals, and more extensive spaces (Thomassen 2018: 91). Laboratories and surgery rooms can be seen as border zones, where liminal lives are disconnected from one body and connected to another. While the body enters these zones as one organism in a particular social and cultural context, another body leaves the laboratory zone to enter a post-liminal area in a biologically mattered form. The liminal laboratory area is a site of struggle for survival, and whether a given liminal being will stay alive until incorporation into another body is not certain. Many liminal beings remain in the liminal forever.

In Squier's vision, borders are emphatically understood not only as physical spots or transgression lines, as suggested by Foucault:

> These liminal lives test the boundaries of our vital taxonomies, whether social, ethical, biological, or economic. As medical interventions are reshaping our way of conceiving, being born, growing, aging, and dying, liminal lives surround us—in our schools, our families, our professions, our institutions, our representations—anywhere that the expected shape or span of human life is being changed through biomedicine. (Squier 2004: 4)

Squier's vision of the constant testing of borders across existential levels ties in with the Deleuzian picture of unstable, continually changing societies of control, to convey which Deleuze and Parnet uses a vivid metaphor of a casting mould losing its shape and a sieve with irregular holes (2007: 184). He stresses that societies of control can never bring anything to completion, and theirs is a never-ending story (2007: 185). Capitalism is perpetually mutating (2007: 186) in the same way as genomes are. Biological and social bodies develop together, sharing the same dynamic. The problem is that there is no moment of closure to the transforming ritual (which was there in the rituals of societies of discipline). In Squier's terms, this entails 'becoming liminal' without any hope of arriving at any post-liminal stage. Discussing the proximity of liminal lives in our daily routines, Squier does not refer to elements of our bodies that have the potential of becoming liminal (being isolated from our organisms). I understand her intention of leaving us in uncertainty. Liminality does not offer certainty.

Liminal laboratory lives, or rather their representations, such as pictures, video recordings, drawings, and descriptions, cross the laboratory borders—the borders of biology/medicine—to reach everybody on a daily basis. They call for attention, showing the mechanism of becoming liminal. Messages from representations of liminal life may easily be passed over or misunderstood. Grasping them is a matter of knowledge and general sensitivity or, as Isabelle Stengers would say, of a public intelligence (2018). Ignoring them can result from a distrust of representations, which are only mediators. Direct contact with liminal lives would supposedly have a stronger effect, but it is impossible because of biosafety protocols and a common fear of elements coming from the body.

Squier discusses an ambiguity lingering in the approach to liminality, similar to that mentioned by Thomassen with regard to post-modern times. Squier avails herself own Paul Rabinow's notion of 'purgatorial anxiety' (Rabinow 1999: 17–8), which denotes a mixture of hope and fear related to technological development. Becoming biologically liminal causes a direct experience of danger. The ritual of moving liminal lives from one body to another is not always entirely successful; it invariably incurs a risk. Importantly, liminal lives sometimes remain petrified in a liminal position (as cells and embryos frozen in laboratory refrigerators). The ritual has no closure (as is the case with social liminality in societies of control). Squire argues that:

> Becoming liminal is a two-way process, in other words: not only do we turn to culture to make sense of moments of biological change, but through our cultural practices—our performances—we are able to access, and indeed to

> produce, a range of meaningful changes in our bodies. The focus on performance can extend beyond embodiment to articulation. (2004: 11)

I would add that becoming liminal is a nature-culture process. Changes in embodiment and in the understanding of it impact changes in our social and cultural existence. The process also works in the other direction, as cultural productions, including literature (especially sci-fi) and performative arts, inspire biological transformations. Scientists can derive inspiration and even obtain meaningful data particularly from the liminal practices of the art&science genre, which encompass experiments done for artistic purposes, such as the coding of graphic forms in bacteria conducted by Joe Davis (Yetisen et al. 2015).

While exploring rituals and liminality, one can hardly avoid addressing pollution and the abject; in any case, they are fundamental to my research and my liminal identity. In my understanding, pollution and the abject are related to contamination. In van Gennep's model of the ritual structure, individuals or a group that undergo a ritual have to leave a particular order of the pre-liminal phase and enter the liminal sphere of disorder so as to finally reach a new post-liminal order. Being 'within the order' and 'outside the order' is associated with what I have labelled as 'not yet' and 'not only.' Liminality is characterized as chaos and the lack of structure. It is a space in which one finds oneself having crossed one border until one crosses another one; in other words, it means existing in the border zone. All aspects of this territory seem to meet the conditions of dirt/pollution and the abject.

Julia Kristeva defines the abject as something 'radically excluded' and drawing toward the point where meaning collapses—something located outside the order (1982: 3). This definition perfectly fits liminal lives and liminal practitioners. Abjects are neither subjects nor objects; in my vocabulary, they are not only/not yet. This is the reason why they are deemed strange or even dangerous. The abject has a robust bodily character, which is crucial for and in becoming liminal and transmattering. Kristeva insists that:

> Abjection preserves what existed in the archaism of pre-objectal relationship, in the immemorial violence with which a body becomes separated from another body in order to be. (1982: 10)

While it is not my goal in this text to analyse the genesis of the abject as a primary affect, Bennett's idea of impersonal affect indicates that the abject can be considered to result from the affectiveness of self-organized matter. Having concluded this, I need to explain why my liminal I-with is abjectual.

The moment of separation from the body is crucial for abjects, as it is crucial for liminal laboratory lives. The examples of the abject marshalled by Kristeva include menstrual blood, body-liquids, cut nails, and lost or cut hair (1982: 3). What they all have in common is that they were appreciated and entirely accepted when still being part of the body. They become disgusting and are rejected when they find themselves beyond the borders of the body, and it is hard to find any function for them. In Kristeva's definition, the corpse is the most significant abject: "The corpse, seen without God and outside of science, is the utmost abjection. It is death infecting life" (1982: 4). The cadaver crosses the essential line between life and death. The rejected abject may seem to have little in common Tsing's notion of contaminated diversity, with the two being ostensibly located at the opposite poles. However, Tsing argues that "contaminated diversity is complicated, often ugly, and humbling" (2015: 39–40), properties that link it to the abject. We accept all aspects of contaminated diversity, different forms and states of matter, including the abjectual ones.

The concept of the abject was meaningful to my research, especially to my fieldwork for the *safe suicide* project. In the course of it, I had to ask myself a series of questions, such as: Are Squire's liminal lives abjectual? Can my cells, derived from my body and placed in laboratory conditions, be perceived as the abject? The point is that liminal lives do not lose their function or value on leaving the body. Rather, they are mattered as a research material or medical items used to cure other bodies. As long as they retain their function and potential, they are part of the laboratory context. This was what happened to my cells, which participated in safe suicide. They contributed to an artistic project, and, as Kristeva notes, religion and art are the areas where the abject can be purified (1982: 17). That would explain why safe suicide as an art&science project prevented my cells from becoming abjectual. However, this protection was not complete. The cell portions that, for one reason or another, went to waste and ended up in the trash left the project context, turning into biohazardous pollution.

The notions of pollution and dirt were axial in my research around *The Last Supper*. The two are closely linked to abjection. As Mary Douglas observes:

> If we can abstract pathogenicity and hygiene from our notion of dirt, we are left with the old definition of dirt as matter out of place. (...) Dirt then, is never a unique, isolated event. Where there is dirt there is system. Dirt is the by-product of a systematic ordering and classification of matter, in so far as ordering involves rejecting inappropriate elements. (1966: 36)

On this model, dirt is something 'outside of a certain order,' something that has been removed as a redundant element. Dirt does not fit in any other order. It pollutes every system. Putting a social or cosmic order at risk triggers an immense ritualistic danger, a danger of the infinite. Douglas avers that:

> Order implies restriction; from all possible materials, a limited selection has been made and from all possible relations a limited set has been used. So disorder by implication is unlimited, no pattern has been realized in it, but its potential for patterning is indefinite. This is why, though we seek to create order, we do not simply condemn disorder. We recognize that it is destructive to existing patterns; also that it has potentiality. It symbolizes both danger and power. (1966: 95)

People tend to create patterns and schemes and to take care of their borders in order to have an illusion of control. These borders also refer to the limits of the body.

The notion of pollution is meaningful in the social discourse on genetic modifications and transgenic organisms. The opponents of genetic mattering stress that such research steps beyond several essential orders, mainly the order of nature, the body order, and the God-created order (Blancke et al. 2015). Transgenic organisms are outside of the biological system in place; they cannot be defined as new species since they do not differ enough from their unmodified prototypes. Consequently, their status is unclear and their relocation outside the laboratory is interpreted as endangering the 'natural' ecological balance. In line with Douglas's reasoning, GMOs (genetically modified organisms) have sparked indefinitely many possibilities, which are difficult both to imagine and to manage. They occupy the liminal laboratory zone, waiting to be eventually incorporated into new natural orders.

The abject and pollution have also been inscribed in my liminal position. Liminal practitioners developing transdisciplinary projects 'pollute' the structured academic system. We are trouble, causing discomfort and enforcing changes in the system. The good thing is that being active in border zones affords us ample opportunity to explore and describe these areas.

The abjectual quality of liminal practices is also related to the production of factishes. As already mentioned, Latour has minted this 'in-between' term to capture what is generated in the laboratory. Of course, scientists are not really eager to agree that they produce factishes rather than facts. For their part, liminal practitioners realize that factishes are the only possible effect of their work, bringing together what scientists believe they produce (facts) and what artists traditionally produce (fetishes). Factishes appear to be abjects of cognitive and creative processes. They are ousted from both fields. This does not mean that they are fixed; on the contrary, their function is to co-vibrate in the border zone.

The liminal abject boasts a robust and stubborn status, being, so to speak, consecrated by its liminality. Rather than fixed in its rejected position, it continually finds itself in productive co-vibration. Neither trash nor even junk, it is simply removed from the disciplinary orders. The liminal abject is an embodied process. The body is neither an object nor the abject; it is a phenomenon in an ongoing process.

## References


Blancke, Stefaan, Frank Van Breusegem, Geert De Jaeger, Johan Braeckman, and Marc Van Montagu. 2015. Fatal Attraction: The Intuitive Appeal of GMO Opposition. *Trends in Plant Science* 20 (7): 414–418. https://doi.org/10.1016/j.tplants.2015.03.011. Accessed 22 June 2022.

Deleuze, Gilles, and Claire Parnet. 2007. *Dialogues*. Revised edition. Translated by Hugh Tomlinson and Barbara Habberjam. New York: Columbia University Press.

Douglas, Mary. 1966. *Purity and Danger*. Durham: Duke University Press.

Knorr-Cetina, Karin. 1991. Epistemic Cultures: Forms of Reason in Science. *History of Political Economy 23 (1)*: 105–122.

Knorr-Cetina, Karin. 1999. Epistemic Cultures: How the Sciences *Make Knowledge*. Harvard University Press, Cambridge, MA

Kristeva, Julia. 1982. *Powers of Horror: An Essay on Abjection*. New York: Columbia University Press.

Latour, Bruno. 1983. Give Me a Laboratory and I will Raise the World. In *Science Observed: Perspectives on the Social Study of Science*, ed. Karin Knorr-Cetina and Michael Mulkay, 141–170. London & Beverly Hills: Sage.

Latour, Bruno. 1993. *We Have Never Been Modern*. Cambridge, MA: Harvard University Press.

Latour, Bruno. 1999. *Pandora's Hope. Essays on the Reality of Science Studies*. Cambridge & Massachusetts & London & England: Harvard University Press.

Latour, Bruno. 2005. *Reassembling the Social. An Introduction to Actor-Network-Theory*. New York: Oxford University Press.

Latour, Bruno. 2009. Dajcie mi laboratorium, a poruszę świat. Translated by Krzysztof Arbiszewski and Łukasz Afeltowicz. Teksty Drugie, 1–2.

Latour, Bruno. 2010. *On the Modern Cult of the Factish Gods*. Durham & London: Duke University Press.

Latour, Bruno. 2013. *Nadzieja Pandory. Eseje o rzeczywistości w studiach nad nauką*. Toruń: Wydawnictwo Naukowe Uniwersytetu Mikołaja Kopernika.

Lynch, Michael. 1985. *Art and Artifact in Laboratory Science: A Study of Shop Work and Shop Talk in Research Laboratory*. London; Melbourne; Henley: Routledge & Kegan Paul.

Rabinow, Paul. 1999. *French DNA. Trouble in Purgatory*. Chicago: University of Chicago Press.

Squier, Susan Merrill. 2004. *Liminal Lives: Imagining the Human at the Frontiers of Biomedicine*. Durham and London: Duke University Press.

Stengers, Isabelle. 2018. *Another Science is Possible: A Manifesto for Slow Science. A Manifesto of Slow Science*. Cambridge: Polite.

Thomassen, Bjørn. 2018. *Liminality and the Modern. Living Through the In-Between*. London & New York: Routledge Taylor & Francis Group.

Tsing, Anna Lowenhaupt. 2015. *The Mushroom at the End of the World: On the Possibility of Life in Capitalist Ruins*. Princeton, NJ: Princeton University Press.

Turner, Victor. 1974. Liminal to Liminoid, in Play, Flow, and Ritual: An Essay in Comparative Symbology. *Rice Institute Pamphlet – Rice University Studies* 60 (3). https://hdl.handle.net/1911/63159. Accessed 22 June 2022.

van Gennep, Arnold. 1960. The Rites of Passage. Translated by Monika B. Vizedom and Gabrielle L. Caffee. Chicago: University of Chicago Press..

Yetisen, Ali K., Joe Davis, Ahmet F. Coskun, George M. Church, and Soek Hyun Yun. 2015. Bioart. *Trends in Biotechnology* 33 (12): 724–734. https://doi.org/10.1016/j.tibtech.2015.09.011. Accessed 10 May 2022.

CHAPTER 4

# Living the Method: Bioart as Autoethnography

I did not find autoethnography; rather, autoethnography found me. I had been practising the autoethnographic approach even before I found out that there was such a method, with my artistic pursuits having always been a form of autoethnography. Autoethnography is a method that does not force anyone to suppress their own identity or interests; it enabled me to be I-with, an artist, a researcher, and an academic at the same time.

At the beginning of my research, I intended to examine the audience of my artistic activity, a theme on which I had started working before joining academia. At the time, I would do it by talking to my public, handing out questionnaires to complete, running workshops, and making them active components of exhibitions in experimental and intuitive artistic production and explorations. However, inexpertly handled quantitative research tools were inappropriate for examining human/human and human/non-human relations. Besides, I realized that being an artist and the originator of those projects and exhibitions, I was susceptible to hermeneutic limitations when analysing the surveys.

The very process of my transmattering that unfolded in the laboratory conditions proved far more important than any assessment of its final effects as put on display at the end of my project. When doing my laboratory work, I was not judged, and I was intra-acting with multiple human and non-human agents. I became a full-fledged participant of a unique social and cultural process, the process of mattering and transmattering.



K. Żyniewicz, *The Body as Matter in Art and Science*,
https://doi.org/10.1007/978-3-032-13665-7_4

Learning about Science and Technology Studies (STS) research on laboratories marked an important turning point as I started to find answers to central questions, such as: What happens when an artist/researcher enters the scientific field? What relations come into being then? What matters in this kind of transdisciplinarity? I realized that I myself was the best tool to use. I needed a methodology that legitimized my deployment of I-with as a voice of the network in and with which I was interconnected. Autoethnography met my needs and expectations.

Participation in laboratory fieldwork is hard to define accurately in my specific case. My lack of biotechnological knowledge prevented my complete involvement in laboratory activities (Complete Member Research). I did not actively contribute to the research carried out by the laboratory team, and I only performed selected tasks instrumental to the implementation of my art&research projects. I did not write research papers with the biologists; nor did I draw up grant applications with them. They came to be accustomed to my presence and work, but my liminal status was perceptible, nonetheless.

On the whole, no research into matter and no examination of transmattering on the biological, social, and cultural levels can possibly take place without taking into account the body-with of the researcher and the bodies of the other human and non-human agents involved. This was also true about my research work, which hinged on an array of embodiment forms: besides the very physical presence in the fieldwork, a transformation—a form of organic transmattering—takes place from the *in vivo* state to the *in vitro* state, from the whole to fragments, from the inwards to the outwards, from death to immortality, and the other way around.

In 2017, I wrote an article titled 'Art&Science: The Autoethnography of a Woman Artist Engaged in Art&Science Projects in Biological Laboratories' (Żyniewicz 2017: 103–26) with a view to explaining why I had decided to embrace analytic, rather than evocative, autoethnography. I wanted to make it clear to everyone, including myself, that I was indeed practising an academic method of research. Then I realized that, if anything, my research activity was aligned with the evocative model.

The results of my art&science practice in biological laboratories expressed in artistic contexts are examples of evocative autoethnographic work. The goal of my projects, which included a substantial component of biological research, was to trigger recipients' emotions and initiate a debate about the significant social and cultural phenomena that were the object of my work. The underlying research was based on my own body, emotions, and vulnerability. The autoethnographic material was rendered in performances, installations, and exhibitions. Every re-presentation was different and unique, manifesting itself in its own way to various audiences. Fuelling unique representations, such cognitive pursuits are the domain of art schools rather than universities and are only reproducible as yet another re-presentation.

Commencing my work in biological laboratories, I began to keep a fieldwork diary as an ethnography of the laboratory from the perspective of I-with. This diary is not supposed to be regarded just as a personal diary that collects impressions and thoughts. It is a diary that documents my experience as an active element operating within a specific network. In my case, the laboratory played the role of an art studio. However, while in the epistemic culture of the laboratory, an experiment has to be repeatable to lend validity to what it is supposed to demonstrate, repeatability was not the central concern in my work. In my research, the scientific experiment was an artistic experiment. During the day, I would take notes recording the new techniques I was learning at the laboratory, and I would edit them after the working hours. Through this regular reflection on my daily experiences, I realized that I existed in at least three different contexts: the biological laboratory, the university, and art.

Scientists—and perhaps other artists—can repeat the laboratory procedures I performed. What is unrepeatable is the idea of the project, the reason for performing those procedures. Given this, it would be a mistake to hold it against me that the experiments I did cannot be repeated and, as such, cannot be counted as proper research. The autoethnographic experiment is, by default, carried out on I-with.

The examination of transmattering reveals uncertainties clustering around the researcher's identity. The researcher and artist in one can be seen as a methodological abject located in the wrong place (Kristeva 1982). As an artist steps into a laboratory, several questions arise, including: What does an artist do in a biological laboratory? What does an artist who works in laboratories do in academia? What does a researcher do in a gallery? These questions explore being an element of networks that are not tightly connected to a given context. It is challenging to a person that belongs to just one of these contexts to move to the other ones, and uncertainty about whether to do it or not lies at the core of experimental liminal practice.

Ethnographers are susceptible to two forms of incertitude or foreignness. One of them is triggered when researchers embark on their fieldwork and need to immerse themselves in a new ecosystem, with their immersion developing, deepening, and expanding over time. The other species of foreignness surfaces when researchers share their results with the academic community, most of whom have not experienced what they went through when doing their fieldwork.

Between the stage of collecting observations and the stage of putting them in writing, an ethnographic moment takes place (Strathern 1999: 1). Strathern, who has put forward the notion of the ethnographic moment, depicts it as linking that which is analysed during observation to "the need to understand (what is observed at the moment of analysis)" (1999: 6).

When doing autoethnography, the researcher must be attentive to the field in which they are acting, not least because they are always required to explain their role and actions in it. Retaining this always-on consciousness can be taxing, especially when autoethnography delves into challenging personal life situations. For example, Elżbieta Zakrzewska-Manterys (1995) dwells on how difficult it was for her to combine the roles of the mother of a child with Down syndrome and a researcher. The ethnographic moment is a border point, and it enables the researcher to move forward, to transform the matter of their research and the matter of themselves.

In the case of my research, a double ethnographic moment and triple foreignness have to be considered since as a result of my triple foreignness (Fig. 4.1), I experienced two ethnographic moments. There was a distinct phase in the laboratory work when the material I had produced was shared in the artistic setting and on the academic stage. Both occasions involved the risk of being unaccepted and considered a foreigner. Co-vibrating was not necessarily smooth, and frictions had transmatterive functions.

The research depicted in this book covered the implementation of two liminal projects in biological laboratories. The connection of these projects to biology is the reason why they are as a rule officially classified as bioart, a term that I find too narrow. I prefer dubbing the two projects as examples of evocative autoethnography or, as Bojana Kunst would put it, the production of a complex artistic subjectivity (2015: 19).

Bioart is an art&science practice founded on a more or less tight collaboration of art, sciences, and technology. This collaborative investment makes art&science a typical capitalistic production of relational aesthetics (Kunst 2015: 53). While various actors from different cultural fields can take part in relations established within art&science practice, biology in all its varieties is the crucial partner of art in bioart.

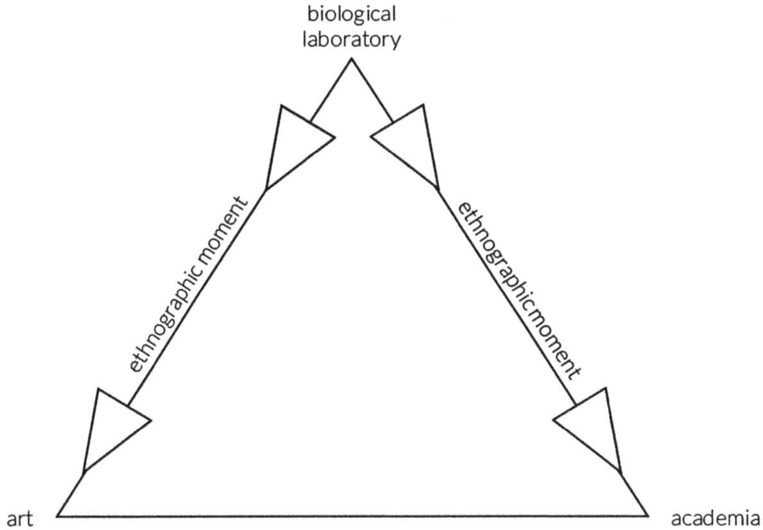


**Fig. 4.1** Double ethnographic moment and triple foreignness

Besides the collaboration of art with the life sciences and the hard sciences, there is also an increasingly more significant interaction between the humanities and art, with creative tools, such as performances and visuals, used together with texts or instead of the text. Importantly, the final products of transdisciplinary/experimental practice at academia must for formal reasons be ascribed to a defined field, such as art, cultural studies, anthropology, sociology, or philosophy, and this imperative fundamentally contravenes the nature of this practice. A pertinent question is how the effects of 'in-between' practices can be retained really 'in between.' To solve the problem for good, an incisive remodelling of the university and research system would be required.

As an additional difficulty in classifying art&science practices, it is hard to list all the forms of its 'in-between' collaborations. Art&science is usually examined in the form of case studies of isolated projects. Thinking of art&science, we think of many different facets of the same cognitive phenomenon in which every artist/researcher develops their methodology and constructs a network of actants. Multiple aspects of the art&science phenomenon also prompt coining many definitions that focus on its different properties. Ryszard W. Kluszczyński, in a paper in the Art+Science Meeting series, observes that:

> (…) art morphs into artistic research and engages in projects aimed at producing diversified knowledge. Within this involvement, I identify four tendencies and four varieties of knowledge: meta-artistic tendencies (reflection on and knowledge about art effected through artistic practice); institutional critique (reflection on and knowledge about the world of art and its institutions); socio-cultural critique (reflection on and knowledge about social orders); and Art&Science or ArtSci (reflection on and knowledge about various areas of scientific and art-scientific research). (2019: 16)

The pivotal point of this passage is that Kluszczyński interprets artistic practice as a species of research. The categories he enumerates can either join forces in the same project or work separately. Often given the moniker of art-based research, art&science activities employ a mixed methodology that can combine seemingly very different disciplines, such as biology, sociology, and psychology. Such blends perfectly exemplify the methodological anarchy extolled by Feyerabend in his famous claim that 'anything goes' as long as it is motivated by a research idea (2010). Artistic research welcomes methodological patchworks and assemblages, which makes a lot of sense, given that art&science is focused on material assemblages, to use Bennett's vocabulary. In this context, art is a way of knitting connections, giving shape to the final result of the research, and channelling its dissemination. Art-based research must be seen as methodologically structured and not as a mere label added to an artefact.

Mitchell has proposed an interesting view of art&science relations, calling them 'ecumenical' and, at the same time explaining that:

> Though the term 'ecumenical' suggests harmony, the process of folding different fields of human endeavour into one another is a fundamentally transformative practice, leading neither to equilibrium nor to ever-flexible relationship of resilience, but rather to a form of transformation better framed in terms of religion than engineering: art enables a ritualistic dying and becoming, rather than bending and flexing. (2017: 74)

This point of view ties in with Kluszczyński's definition by foregrounding the aspect of transformation, which means that the partnership of arts and sciences is not a passive contiguity but a tectonic shaping of each other. This supports my concept of co-vibrating agents distributed across different fields. At the same time, Mitchell's ecumenical vision, though promising, may be overly idealistic. It entails the agreement of all the participants in art&science disciplinary relations to produce factishes. My experience suggests, however, that factish production may be a goal to achieve for art, but abandoning fact production is still a steep challenge for sciences. I have found that smooth co-vibrating is possible until the assessment of project results. Nevertheless, Mitchell is right to claim that art brings a ritualistic atmosphere to science, or, to put it more precisely, that art brings out the ceremonial character of science production. This ritualistic character of art is also related to the capitalistic tendency to display performance/production instead of the final product (Kunst 2015).

Another image of art&science practice has emerged from the observations of artists carried out by other artists under the residency programme Artists-in-Labs, in which various artists visited selected laboratories. Culminating in the compilation of their experiences and notes, the venture can be interpreted as the subjectification of artists by focusing attention on an artist in action (imitating Latour's coinage, 'an artant' would be a suitable term to capture this). Introducing the collected volume on Artists-in-Labs, Jill Scott explains that:

> Therefore, the artist's reports in this book are part of a larger 'narrative ethnography' about how the artists spent their time in the various labs. As a research team, our observation in the Artists-in-Labs project was from the stance of field sociologists, in other words, our attempt was to observe and study a social scene and whenever possible, we tried not to interfere with the situation. (2006: 6)

The Artists-in-Labs residency publications offer insights into the social aspect of art&science practice. Attention to this factor is no novelty as such. For one, as early as the 1930s Walter Benjamin (1970) pictured artists as producers reflecting on the political and social conjunctures at hand. All the participant of the residency programme shared their experiences and observations, which, when assembled, rendered a panorama of relations between actors from arts and sciences. The two Artists-in-Labs books show what can be problematic about multimethodologies, what advantages they provide, and what the majority of art&science collaborations have in common. One conclusion formulated in the Artists-in-Labs publications is that many problems result from the tendency of science and art to marginalize each other (Scott 2010: 8). This mutual deprecation is engendered by the social perceptions of different fields. Science appears to be more respected and better funded. Establishing partnerships is thwarted by mental barriers, and if scientists view artists as lesser and less important, they may fail to see how collaboration with artists can benefit them. At the same time:

> Artists should not enter a science lab to use the lab tools or to 'be a parasite' but to see scientific machines as metaphors for all kinds of systems, explore scientific theories as compositional strategies, and not ignore the disturbances and imbalances, mutations and reproductions, and errors that may be possible. It would indeed be sad if they became scientists, directly transferring methods, copying their orders and innovation processes. (Scott 2010: 10)

Stepping into laboratories, artists obtain an opportunity to observe science production as laboratory ethnographers. Such involvement helps artists to understand what is happening in science production. If artists are not bound to be uniformly appreciative of what is going on, and their perception can of course be critical, being critical about the collaborators in a liminal project is an iffy thing to do. The development and the implementation of the project usually take long enough for people to make friends with each other. Criticizing one's friends in a public exhibition can feel like betrayal. Besides, art&science practices, like all of contemporary art, is a sector of capitalistic production, which means that, besides artists/researchers and scientists, other actors are also involved in the undertaking, such as sponsors, funders, and cultural institutions. To make the development and exhibition of a project possible, its very idea must inevitably be negotiated, and a part of it must often be surrendered in one way or another. Like any other domain of capitalism, artistic ventures are project-based (Kunst 2015: 153). To obtain funding for a politically risky project is far from easy, if indeed possible in the first place. Therefore, relationality has two facets. It affords a chance to execute complex and exciting projects, but it also requires taking every participant into account, which can preclude some solutions.

Many misunderstandings about art&science practices result from the notion that they are merely a form of science popularization, which presupposes that artists cannot be discursive and that they should only be illustrative and affirmative. Many scientific laboratories invite—or at least welcome—artists as participants in the belief that such joint enterprises will be an excellent way to advertise their own research. Meanwhile, the possibility of more profound reflection and more meaningful dialogue is crucial for liminal practices. Both parties involved, artists and scientists, need to exhibit mutual respect and understanding, collaborating instead of exploiting each other. As all research and artistic fields each have their own official or unwritten code of professional ethics, liminal methodologies also boast an ethics of their own. This ethics can be understood as cultivating our methods and endorsing the methods of the area we want to explore. Art translates the scientific language into its characteristic language, extending the translation chain that unfolds in science production.

Another critical function of art in its interaction with science is to debunk the myth of objective science. Even though the objectivity of science has already been interrogated and undermined, this mythology of science is still largely present in social perception:

> When the artists entered the science laboratories they automatically started to transform and mediate the image of the exact and objective science. With a kind of 'shamanistic unsharpness' they turned scientific artefacts into experiential realities and interpretations: in other words, they represented science in a different way. As David Bohm has suggested 'Representation' is in fact the most appropriate term for this transformation, as it means 'to re-present' again. It is my conviction that interdisciplinary dialogues require a new level of profound reflection, and we should risk asking important questions which reach beyond the boundaries of scientific objectivism. (Stettler 2006: 42)

The 're-presenting' of science appears to be the key point here. Artists re-present scientific representations, and they translate them into the language of art. This language is not necessarily easier to learn and understand. With all its richness, the language of art is undoubtedly capable of bringing out aspects of reality and science that would otherwise go unnoticed. For its own part, the language of art&science practices can be so complex as to require a parallel narrative explanation. Explaining, which can also be classified as storytelling, is critical in art&science activities.

The importance of art&science narrative has been highlighted by Thierry Bardini. His *Junkware* examines two well-known bioart projects—*Microvenus* by Joe Davis and *Genesis* by Eduardo Kac—to conclude that contemporary art should be suitably called junk art and to note that "on an interesting new twist on the history of art, the visible has given way to the readable, (...) the aesthetic posture requires the explanatory discourse of the artist" (Bardini 2011: 203). Practices combining the languages of science and art are also challenging in terms of production and reception. Both projects discussed by Bardini were predicated on the genetic modification of bacteria. This transformation does not cause any visual effects. Even looking at the bacteria through a microscope does not enable one to see the outcome of the artistic operation performed with them. In this case, the aesthetic lies in the decisions the two artists made to conduct meaningful experiments and in the process of their implementation. The only way to prove that Davis and Kac did what they said they had done is to follow their scientific protocols to investigate the changes in the bacterial genome.

These two projects also perfectly exemplify the adjustment of contemporary art to the rules of capitalism. They are both based on materialized speculation. As Marina Vishmidt has argued, speculation is the most vital mode of production today (2018). Creative transformations of life are the pillars of capital. However, creation is by no means the monopoly of art. In general, the dividing lines between all spheres of activity in life have become blurred. Kunst observes that today's art is based on demonstrating processes, sharing experiences, and performing. Artistic performances produce life. The theme of the production of life directly leads to bioart. It is not my goal here to describe the history and characteristic of the field. However, because the fieldwork described in this book is based on the development and implementation of my two projects in biological laboratories, my position on and definition of bioart must be explained. I decided not to call my practice bioart and to remain in a liminal place instead.

The approach espoused by the Centre of Imaginative Ethnography has helped me to understand my art&research practice as a form of ethnography. Given that the projects I analyse here were based on biomatter derived from my own body, I regard them, more specifically, as a form of evocative autoethnography. I find this way of defining far more convincing than stating that in bioart "many contemporary anthropologists and artists are turning their attention to creative methodologies, to articulating ethnography and artistic practices in the process of research, and to ethnographic knowledge co-creation" (Culhane 2017: 7).

Creative methodological approaches blur the lines separating art from sciences and the humanities and grant one freedom to shape experimental cognitive processes. Continuing Culhane's thought, the keywords for contemporary anthropology and ethnography include dialogue, collaboration, ethical engagement, imagination, and creativity. These values are evoked in the definitions and depictions of art&science and bioart above.

My art&science activity perfectly fits Kunst's idea of the salience of artistic subjectivity. Kunst notes that the second half of the twentieth century was marked by a crisis of subjectivity. Therefore, regaining subjectivity became the supreme goal of creative practices (2015: 20). Not all bioart projects are dedicated to the artist's subjectivity. Artists/researchers do not always perform using their bodies or life in general. Many projects are devoted to non-human living beings and being-with them. I decided to work with my own body and other bodies isolated from mine and to analyse the totality of my experience of and with biological/laboratory operations. My social roles as an artist and a researcher have fused into one, engaging me fully in production and selling.

All definitions of experimental ethnography and evocative autoethnography are convergent with my outline of art&science and bioart. All these experimental forms of production can be considered creative methodologies, which should "facilitate our communication with diverse audiences both in the academy and out" (Culhane 2017: 12). In my project, my method has had to facilitate communication with people from academia and the artworld. My art&science activity can be labelled as performative ethnography, which demonstrates that the same pursuit can unfold in and be associated with various contexts.

Magdalena Kazubowski-Houston states that there are two different modes of doing performative ethnography:

> I practice what can be referred to as 'performance of ethnography' (performance as ethnographic representation) and 'performance as ethnography' (performance as ethnographic process). Performance and ethnography can be combined in various ways. One can conduct research and then develop a performance based on it; use a performance development process as a way of conducting fieldwork; or use performance as an ethnographic process that culminates in a publicly staged performance. (2017: 114)

The two models identified by Kazubowski-Houston were combined in my research. *The Last Supper* and *safe suicide* were performative projects. However, selling them in the academic setting and in the artistic setting took different trajectories.

## References

Bardini, Thierry. 2011. *Junkware*. Minneapolis, London: University of Minnesota Press.

Benjamin, Walter. 1970. The Author as Producer. *New Left Review* 1 (62): 1–9.

Culhane, Dara. 2017. Imagining: An Introduction. In *A Different Kind of Ethnography: Imaginative Practices and Creative Methodologies*, ed. Danielle Elliott and Dara Culhane, 1–23. Toronto: University of Toronto Press.

Feyerabend, Paul. 2010. *Against Method: Outline of an Anarchistic Theory of Knowledge*. New York: Verso.

Kazubowski-Houston, Magdalena. 2017. Performing. In *A Different Kind of Ethnography: Imaginative Practices and Creative Methodologies*, ed. Diana Elliott and Dara Culhane, 113–133. Toronto: University of TorontoPress.

Kluszczyński, W. Ryszard. 2019. On the Boundarylessness of Art. In *Beyond Borders – Processed Body – Expanded Brain – Distributed Agency*, ed. Ryszard W. Kluszczyński, 6–20. Gdańsk and Łódź: Wydawnictwo Uniwersytetu Gdańskiego, CSW Łaźnia.

Kristeva, Julia. 1982. *Powers of Horror: An Essay on Abjection*. New York: Columbia University Press.

Kunst, Bojana. 2015. *Artist at Work, Proximity of Art and Capitalism*. Winchester, Washington: Zero Books.

Mitchell, Robert. 2017. Smart Art: What Is Art, and What Could Art Be in the Age of Smartness? In *Die and Become! Art and Science as the Conjectured Possible*, ed. Dmitry Bulatov, 561–761. Gdańsk and Łódź: CSW Łaźnia.

Scott, Jill. 2006. Introduction: Processes of Inquiry. In *Artists-In-Labs Processes of Inquiry*, ed. Jill Scott, 6–36. Springer-Verlag/Wien & HGK Zürich.

Scott, Jill. 2010. Introduction: Networking Is Both an Art and a Science! In *Artists-In-Labs: Networking in the Margins*, ed. Jill Scott, 8–11. Wien: Springer-Verlag.

Stettler, René. 2006. Perception-Translation-Transformation. In *Artists-In-Labs: Processes of Inquiry*, ed. Jill Scott, 36–43. Vienna: Springer-Verlag/Wien and HGK Zürich.

Strathern, Marilyn. 1999. *Property, Substance and Effect. Anthropological Essays on Persons and Things*. London & Brunswick: The Athlone Press.

Vishmidt, Marina. 2018. *Speculation as a Mode of Production*. Brill: *Forms of Value Subjectivity in Art and Capital*.

Zakrzewska-Manterys, Elżbieta. 1995. *Down i zespół wątpliwości. Studioum z socjologii cierpienia*. Warsaw: Semper.

Żyniewicz, Karolina. 2017. Art & Science. Autoetnografia artystki realizującej projekty w laboratoriach biologicznych. *Kultura i Społeczeństwo* 61 (3): 103–126.

CHAPTER 5

# Human Liminal Being in Biological Laboratory

My autoethnographic work is an experience, storytelling, and the sharing of a narrative. Every story is a process extended over time. My job is to reconstruct the entire process in my memory and reproduce it in writing and/or performances, installations, and objects. It is like creating a theatre play. The play must be transferred from the theatre of the laboratory to that of the gallery to enter the art context. The challenging part is that as a liminal practitioner I must experiment with I-with, composing a review of its actions. The process of writing and exhibiting is as essential as the laboratory actions and the inquiry preceding it.

While autoethnographic research as such is not necessarily a linear experience, a written account of it must have a discernible order. In my narrative, this order is based on the notion of trajectory, which I understand as a story extended over time. The very term 'trajectory' comes from the grounded theory, where it was first used to describe dying as a border ritual (Glaser and Strauss 1974). Later, the meaning of 'trajectory' was expanded to cover any situation generating suffering, even if not involving the death of the suffering subject (Riemann and Schütze 1992). In this revised understanding, the trajectory is more like a ritual of passage than like a border ritual. In this sense, the trajectory proved helpful in my research as a tool for analysing some periods in life which are not definitively closed by death (Berger et al. 1974).



K. Żyniewicz, *The Body as Matter in Art and Science*,
https://doi.org/10.1007/978-3-032-13665-7_5

A trajectory is best viewed as a portion of autobiography that is meaningful in the life of the person who writes it (the researcher) and, as such, it can be regarded as life in life (Zakrzewska-Manterys 1995). In my research, trajectories unfolded in the development of two art&science projects: *safe suicide* and *The Last Supper*. In both of them, the trajectories included three context-specific levels: the laboratory, the gallery, and academia.

In my view, exhibiting and autoethnographic writing are based on the reconstruction and demonstration of the trajectory. A project can be exhibited and discursively analysed many years after its completion, when the artist/researcher is involved in a different trajectory. Every demonstration has a different influence on its recipients (exhibition audience or readers).

In this chapter, I will focus on my laboratory trajectory. The trajectory of producing *safe suicide* and *The Last Supper* was multi-level and multidimensional transmattering taking place in a biological laboratory, balancing between living and dying and between *in vitro* and *in vivo* conditions. The course and circumstances of this trajectory were bound up with the laboratory scenario and its enactment that formed a transmatterive trajectory setup. Transmattering did not end when the trajectory was completed; it has continued through several demonstrations.

My autoethnographic text is based on excerpts from the fieldwork journal I kept when working at several laboratories between 2016 and 2020. Studied in detail by Science and Technology Studies (STS) scholars, the laboratory reality of knowledge production is explored here from the perspective of a liminal human being (who will also be called a liminal practitioner, a liminal researcher, and an artist/researcher).

My fieldwork journal used in this book is entirely devoted to the two projects, whose ideas determined all my decisions and laid the ground for all the relations I established. Initially, the concepts behind the projects were rather abstract, surreal, and highly speculative, but they were also perfect triggers of experimentation and discussions. The artistic nature of my projects is what sets me apart from STS scholars. Latour first joined the laboratory as an academic, and then he obtained a technician's position to get his hands 'wet' and understand laboratory work as an embodied experience. The way the 'outsider' is introduced on entering the lab largely determines their later relations with the other actors. When I joined a biological laboratory, I had to specify that I was an artist and an ethnographer and that I had some hypotheses to test. Being a PhD candidate was my additional position which lent me certain credibility. Besides my status, I also had to explain the idea of liminal projects to the lab biologists. They needed to understand the main goals and outcomes I was planning to achieve, such as artistic exhibitions, conference presentations, research papers and other publications. Importantly, at a certain point, I was working on the two projects simultaneously, so their laboratory trajectories overlapped.

Contrary to what its title may suggest, *safe suicide* was not dictated by theories of suicide and death—or any other theories for that matter. Rather, it was informed by a very personal concept of a symbolic annihilation of I-with, without crossing the outer border of the body. Theories that bore relevance to my undertaking only appeared in my research much later. For instance, in terms of Durkheim's suicide classification (1987), I wanted to commit egoistic suicide or, more specifically, I wanted my cells to commit egoistic suicide. Later on, in the phase of laboratory production, I realized that my suicidal actions were both active and passive since my cells were dying because of poisoning and the lack of nutrition. That made me ponder whether my cells were committing suicide or whether I was killing them. Before consulting the project idea with the biologists, I was not certain if my vision could be materialized. The only thing I knew was that *in vitro* conditions would enable me to work by delegating my cells to commit symbolic suicide on my behalf. As indicated by the title of the project, my main goal was to do something radical and safe at the same time. It can be regarded as a perfect example of limivoid experience.

The other meanings of the project arose later, when I was more aware of what I had to do in the laboratory. At a certain point in my research, I was able to step beyond my initial position (I wanted to find a way to kill myself without harming my body) and integrate my play with a broader discourse on technologically mediated transformations of the materiality, signification, and concepts of the body. Following the many years of my art&science practice dedicated to life and death, I wanted to play with these notions in the laboratory setting. I did not even know what it exactly was that I wanted to extract from my body; the only thing I knew with any certainty was that I wanted to kill it or make it kill itself (which could take an array of forms in the laboratory conditions). At that time, I had no idea that there was something called apoptosis (the suicidal death of a cell) or that it was possible to make cells immortal. Derived from the lore and terminology of biology, these notions made the project more complex. Killing immortal cells seemed to send a much stronger message than killing a regular cell line. Additionally, I could play with this notion in the laboratory and outside it.

Yet new layers were added to my concept when I came across the story of HeLa cells and their donor Henrietta Lacks. Finding out about those events influenced my work on the project, raising questions about the ownership of the body and the subjectification of cells.

Eventually, the collective decision made in the laboratory was that I would work with my B lymphocytes to create a cell line and experiment on it, starting from immortalizing the cells.

Unexpectedly, when I had already spent more than one year working on this project, an opportunity appeared to develop it in new ways, based on cellular ageing research. While these prospects could promote a separate project, I decided to incorporate the work with the ageing of my cells into *safe suicide* because it also offered an insight into life and the passage of time finalized in death. Cellular ageing, investigated through the colouring of cells, worked as a specific temporality of the trajectory. Interestingly, it did not overlap with the temporality of the project. That made me think that I-with was embedded in many different temporalities which co-created a given trajectory.

Examining ageing, I worked with my skin cells: fibroblasts. To isolate them, I had to have a skin biopsy. Investigating cellular ageing and cellular death included different scientific data and visual representations. They are interwoven in the academic context, but they do not necessarily share the same gallery trajectories.

*The Last Supper* and the *safe suicide* were twin productions because both started from my body (I-with's ingredient) and their implementations coincided in time. *The Last Supper* was scientifically grounded in genetics; technically, however, I had to move from one laboratory room to another to work on the project. I decided to work with my DNA, isolated from the B lymphocytes that were used in *safe suicide*. My major idea was to play with the notion of a miracle, which is usually associated with religion. I envisioned the gallery trajectory of *The Last Supper* as a ritualistic experience, an on-camera performance with the audience playing the roles of apostles. Importantly, the project in many senses exemplified art as a capitalistic production, to use Kunst's formulation, given that the plan was to sell action and engage the audience as much as possible. The recording of the performative supper has since worked as a representation and a memory of the trajectory. If screening it as a video does not engage the audience at all, the action component is still retained.

Besides carrying out a personal operation on I-with, I wanted to inquire whether technology—in this case, biotechnology—would soon enable humans to perform miracles or whether it was only an illusion. The project was also in dialogue with the Christian tradition and art history, primarily Leonardo da Vinci's famous fresco *The Last Supper*. The gallery play was conceived of as a re-enactment of the scene depicted in the painting. Specifically, it presents the biblical moment of transubstantiation, that is, the transformation of wine and bread into Christ's blood and flesh. My idea was to use genetics to perform a similar act. Together with my lab collaborator, I extracted one gene—GAPDH—from my DNA in order to transfer it to yeast cells and in this way to make it participate in the fermentation process caused by yeasts. In my project, the alcohol produced through fermentation was beer instead of wine. This difference underscored the fact that while I had been inspired by a biblical episode, I had actually composed a new script. In the end, the transformed GMO yeast was also used for baking bread snacks, which were served with the beer during a real dinner performance held as part of the laboratory scenario, in a setting reminiscent of da Vinci's painting.

The content of this chapter can be regarded as a regular laboratory ethnography reporting observations of the laboratory world from a beginner in wet work. The trajectory shows how I was learning laboratory work and constructing relations with the scientists. The chapter shows what had to be done to make transmatterive experiments possible. Without charting these circumstances, the transmatterive processes would dangle in a void. In this chapter, I do not address my relations with post-human liminal beings, which have been crucial for transmattering. The laboratory trajectory pictures the production of knowledge and can be of interest to STS researchers.

As already explained, my liminality was founded on my constant co-vibrating with other actors in the structured fields of the laboratory, academia and art. Co-vibration as such allows becoming liminal continuously, and it may be regarded as an incomplete rite. To determine whether liminality is a methodological choice or a natural predisposition is difficult since the two tend to overlap, but this distinction ultimately is not essential. The essential thing is one's incessant cognitive arousal and urge to experiment.

To develop my two liminal projects, I needed to spend some time co-vibrating with actants in biological laboratories. From the physical point of view, laboratories were spaces of my habitat and fieldwork area. Mentally and cognitively, I never left the liminal zone; I just channelled my vibrations towards biotechnology. My entry into the realm of laboratories was a natural course of events. After years of being involved in projects dedicated to the body, I proceeded to the Do It Yourself (DIY) laboratory stage and produced ephemeral living objects. Subsequently, I worked for a long time with insects, exploring the cultural concept of the 'worm/bug' and engaging in bio-sociological experiments. Before embarking on the *safe suicide* project in a regular biotechnological laboratory, I did an internship in the laboratory at the Copernicus Science Centre, where I drew on my observations of burying beetles (*Nicrophorus vespilloides*) and blowflies (*Calliphoridae*) to develop prototypes of living exhibits for exploring the theme of the circle of matter. I learned to run and control the reproduction and transformation of those insects, even in winter, the period of diapause. During this internship, I worked in a sterile laminar flow cabinet, dealing with the *in vitro* reproduction of sundew. While working with insects, I was in touch with and regularly consulted several entomologists (from the Museum and Institute of Zoology of the Polish Academy of Sciences, the Malopolska Centre of Biotechnology, the Faculty of Biology and Veterinary Sciences at the Nicolaus Copernicus University, and the Department of Animal Physiology and Developmental Biology at the Adam Mickiewicz University), but these contacts did not entail working in their laboratories. It took me a long time to obtain access to biological laboratories and eventually 'get my hands wet.' The fulfilment of my effort coincided with my enrolment in a PhD programme and, consequently, the adoption of the additional role of an academic researcher (the academic dimension of the trajectory). The Institute of Genetics and Biotechnology at the University of Warsaw's Faculty of Biology became my primary workplace and habitat.

As soon as I started working at the laboratory, I began to note down everything that was happening day by day: what I had learned, what I had seen, what I had understood. I was experiencing a unique cognitive process that I was convinced had to be described and shared.

I first entered the Institute of Genetics and Biotechnology (my primary habitat, as already indicated) on 16 February 2016, when I came over to meet the team and the people who would help me with my safe suicide project:

> After talking with me in his office, Professor G. showed me around the laboratories, introducing me to the staff and asking them to let me watch their

> work and be ready to answer my questions. Most of the people there were very surprised that I would work on an artistic project at the lab. They were equally surprised to hear that I was working on a PhD with the professor as my advisor. (16 February 2016)

The first reaction of the people I met at the Institute was quite natural. They were faced with a script that looked like one taken from a different theatre. It was like a business meeting or a grant interview. I had to sell my abstractive ideas to them and convince everybody that it was worth investing their resources in enabling me to produce our joint performance. The difference was that in this case the capital was not money. It was what Kunst and Vishmidt refer to as a value. A successful deal would make it possible to me to start my laboratory trajectory. Later on, I had to go through similar negotiations with workers of art institutions to construct the gallery trajectories. Professor Golik was the most decisive person in the situation. He was my gatekeeper, who allowed me to enter the laboratory reality. His acceptance of the value that I proposed was essential. In the further stages of my laboratory trajectory, he also helped me to find new scientific collaborators by putting me in touch with his colleagues from various laboratories. They also had to buy the value that I offered.

The biologists who signed up to co-work with me did not understand all the details of my script at first: What did I mean when talking about committing safe suicide in the laboratory? What did cell culturing have in common with suicide? Nevertheless, their complete understanding was not necessary. It was enough for them to know that they were supposed to help me to isolate my cells and start a cell line. It took a while before they mustered courage to admit how sceptical they had been about my presence in the laboratory. They only told me that when they had come to feel safe and comfortable in my presence—that is, when I was more or less settled. The co-vibrating, so to speak, became smooth.

The Institute consists of five laboratories. I closely collaborated with two of them. Yet, I needed to be introduced to all the staff in order to prevent them from being confused by my presence. Because my first project, *safe suicide*, involved working with my own cells, people from the laboratory researching mitochondrial diseases (mainly Leber hereditary optic neuropathy) agreed to mentor me. I began to visit the laboratory regularly (almost every day) to learn cell culture. Later on, I only sporadically came in touch with some of the people I had met on that first day, and hardly ever with others. The sterile work area seemed to be mysterious and not easily approachable. Not everybody was allowed to enter it—only people working with human cells. The atmosphere of this room produced a peculiar synaesthetic experience:

> The cell cultivating area is a small and cramped room, drenched in the smell of ethanol. The entrance has double doors secured with a lock (it is impossible to open the second door before the first is closed). To work there, one must wear a lab coat and special shoes, only reserved to working there. The general rule is that whatever is leaving the sterile room is not allowed to return. There are two laminar flow cabinets located next to each other in it. There is also a simple microscope for observing the cultures and counting the cells there. The room is so tiny that only two people can work there at the same time. At the beginning, I was terrified to touch all the expensive laboratory devices. I needed to learn the routine of this work. (1 March 2016) (Fig. 5.1)

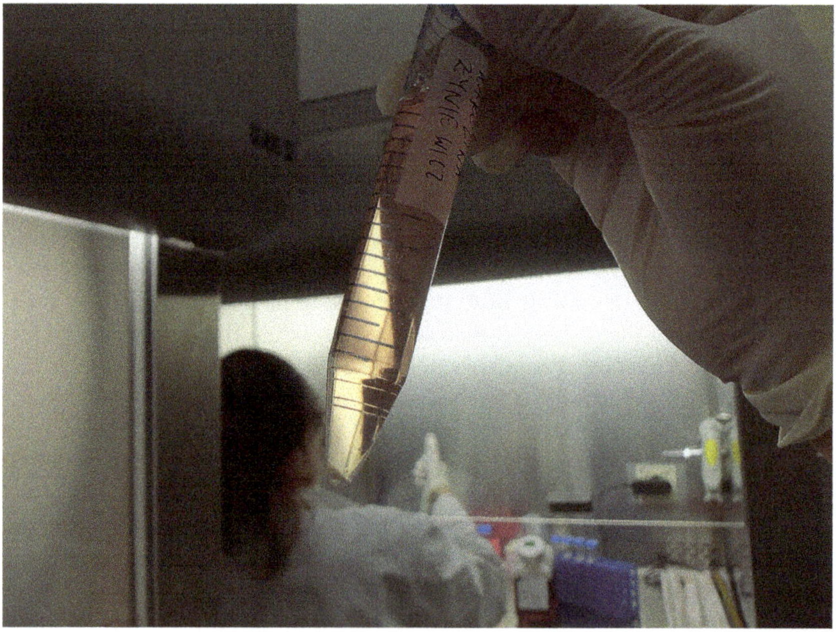

**Fig. 5.1** The cell cultivating area (laminar hood). Photo: Karolina Żyniewicz

Fundamentally, the laboratory environment I entered was a peculiar place of its own kind. The sterile workroom was almost like a ceremonial venue, out of bounds for most people. Having access to this space was evidence that I had sold my idea successfully and gained trust.

Throughout my career, I have worked at five laboratories dedicated to cell culture experiments. Each of them had a slightly different set of rules. Only the one at the Institute of Genetics and Biotechnology described above had a separate room for sterile work. Not all of the labs required wearing a lab coat and shoes reserved exclusively for the laboratory work. What was similar in each sterile work area was that they observed the order of work based on biohazard rules:

1. Before starting work, one is to switch on the UV in the laminar flow cabinet (for 10–15 min.) and then clean the laminar flow cabinet on the inside with ethanol (70%). The last preparatory thing is to switch the regular light and the air circulation back on.
2. One must wear disposable gloves. Working in sterile conditions, a person must not touch anything inside the laminar flow cabinet with their bare hands. It is good practice to spray one's hands with ethanol.
3. Laboratory plastic items are disposable. A new pipette must be used for every transfer of cells from flask to flask and for every substance administration
4. One must not touch a pipette, even if wearing gloves, especially not the end that comes into contact with the medium with cells. The covering paper can be removed from the pipette only when already connected to the pipettor.
5. In the event of randomly touching a pipette, one needs to replace it.
6. Cultivating containers (culture bottles, Petri dishes, etc.) should remain open as briefly as possible; the best practice is to open them only for the moment of pipetting.
7. Optimally, everybody should have separate culturing mediums and reagents, with one's name and the date of first use clearly marked on the label.
8. Before changing the medium in cell culture, the medium bottle must be moved from the refrigerator to the water bath (with warm water). If the medium is too cold, it can cause a shock in cells. The temperature of the medium bottle is tested by touch (like the

temperature of a baby's milk bottle). When the bottom of the bottle is warm, the medium is ready to be administered to cells.

9. If working with blood (cell isolation), one must carefully clean the laminar flow cabinet and the whole room. Everything must be cleaned with ethanol (70%). There is also a hospital disinfectant for cleaning the floor.
10. There are different trash containers. Everything that contains biomatter is a biohazard and must be autoclaved (disinfected at high temperature). Paper towels, gloves, and pipette covers must be disposed into separate trash containers.
11. If an item in the laminar flow cabinet is running out (pipette tips, Pasteur pipettes, etc.), one must replace the package to prepare the workstation for the following user.
12. Having finished the work, one is to clean the laminar flow cabinet on the inside with ethanol, close the window, and switch the UV on (for 10–15 min.). It is essential to switch the UV off after sterilizing.
13. Planning the work for the following day, one must put one's name in the work planner to book the laminar flow hood.
14. All those that use the sterile workroom take turns in the cleaning of it. There is a schedule to follow. (8 September 2016)

As described above, all activities in biological laboratories are based on repetitive protocols, which I reproduced in my notes, enumerating each step. Interestingly,

I transferred the tendency to draw up lists onto the writing of research papers in the humanities. The laboratory practice is performative, but it is worlds apart from free creative pursuits. It can be depicted as a struggle to have things, or rather, at least something, under control.

> For a person from the field of art who has got used to doing everything in their own style, following the laboratory discipline can be challenging at first. (3 March 2016)

After learning and observing the way the scientists worked with cell cultures, I had my blood taken to isolate B lymphocytes from it. I was to work with these cells because this was aligned with the research of my scientific mentors. Professor Golik was my general advisor, but I worked on a daily basis with one postdoctoral employee. Her PhD dissertation, which in part studied the immortalization and apoptosis of cells, was the primary biological reference for my *safe suicide* project. Significantly, the speculative vision I came up with at the beginning was pushed further by the laboratory circumstances. Nothing would have been developed if I had stubbornly clung to my initial imaginations.

Blood taking was the pivotal moment of my laboratory trajectory. Only one person at the Institute was qualified to take blood samples, so making an appointment hinged on convenient timing. Besides arranging this practicality, my other vital need was, as always, to collect evidence: pictures, short videos, sketches, and scientific papers. As Kunst points out, documentation is the only way to sell processes inaccessible to the audience. The moment of blood taking was symbolic for my project, as it marked the separation of my cells from my body (I will analyse it later as one of the transmatterive moments). Therefore, I needed someone to help me to take pictures:

> Some of those present were mildly surprised that I watched the process of the needle being inserted and my blood filling the syringe. Another surprise was that I asked to take photos because I needed to document each stage of the project. Many people teased me about it, which made it very difficult for me. (15 March 2016) (Fig. 5.2)

In a sense, I was ashamed of this preoccupation with documenting. Scientists do not need to sell the process, and they tend to sell the effects of laboratory processes. The laboratory theatre puts plays on without external spectators. Even team members do not watch each other. With time, attention to documentation as part of my practice ceased to be surprising to the biologists, who would often help me with it. Sometimes, one or another person reminded me that a given activity should be recorded. I would say that the biologists became the audience of my theatre performance.

The second stage of the *safe suicide* project started with another small-scale initiation. I entered a new laboratory affiliated with a different institute. To start with, biomatter was again taken from my body. This time, I went through a skin biopsy, and problems with documenting the procedure appeared again. I was sure that it was a potentially transmatterive moment, and I knew I had to capture it, but taking pictures seemed impossible for many reasons.

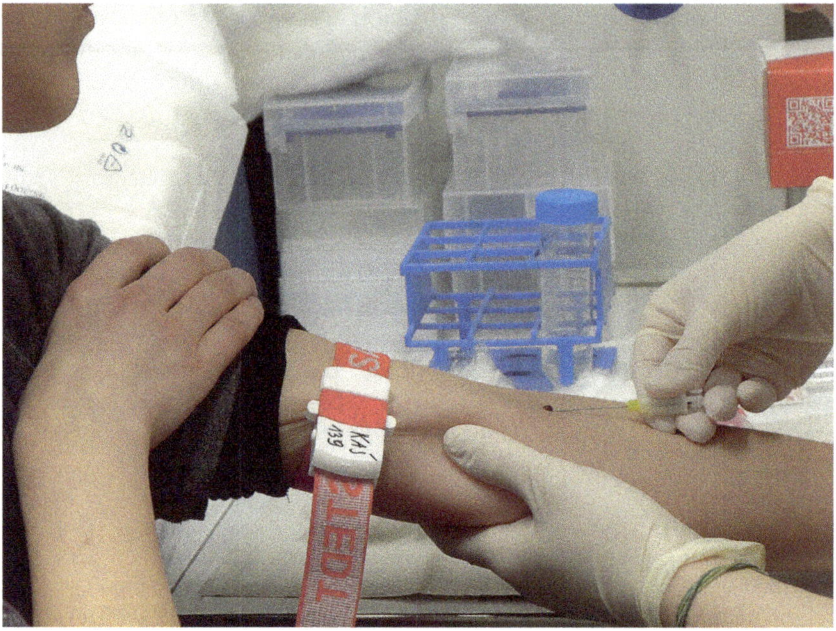

**Fig. 5.2** The moment of blood taking. Photo: From Karolina Żyniewicz's archive

Thanks to A., I got in touch with a doctor who agreed to do the biopsy of my skin. She recommended buying a particular anaesthetic patch and placing it on my arm a few hours before the procedure. I also had to bring a Falcon tube with PBS and antibiotics to transport the biopsy to the laboratory. Going to our meeting, I forgot to take that Falcon tube. Fortunately, the institute was close enough to my primary lab to run back and take it. The doctor had scalpels and everything else one needs for a biopsy. She asked me why I needed to have this biopsy, and the reason was surprising to her.

I was ashamed to mention that I would need pictures, especially that there was nobody there to take them. She was cutting my skin, and I was being cut, so we both were busy. I glued the patch at an ill-chosen spot, just above the elbow, on the outside of my arm. That meant that the trace of the biopsy would be visible for ever. The doctor asked me how big a piece of my skin I needed. I was not sure about it. The anaesthetic patch worked perfectly. I did not feel anything. I was not afraid of being cut. The doctor used pliers to hold my flesh, and she cut a small piece of my skin. I was watching what she was doing, which surprised her because, typically, people do not

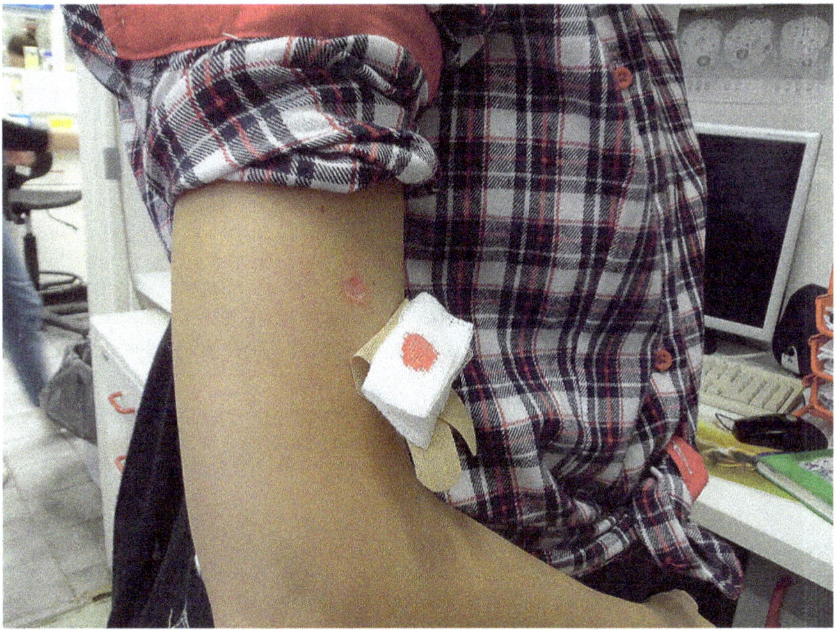

**Fig. 5.3** The wound from my skin biopsy. Photo: Magdalena Kaliszewska

like watching medical procedures. I regretted that I was unable to take any pictures. (24 October 2016) (Fig. 5.3)

The only documentation I managed to produce was at my home laboratory (IGiB), where people were familiar with my need to capture whatever, I did. Without evidence, I would not be able to sell the laboratory process on the artistic and academic markets. Fibroblast isolation was a new challenge and an opportunity to learn something new. I was curious if working with fibroblasts would be different from working with B lymphocytes, and it somehow restored the initial excitement I had felt when first embarking on laboratory work.

I needed to isolate fibroblasts on the biopsy day because fresh biopsy cannot be stored for long. The same person who was helping me to work with B lymphocytes agreed to help me with my skin cells. She divided the skin piece into two parts. One was for her to show me the work, and the other was for me to work with. Working with such a tiny bit was tricky. First, I had to separate the epidermis from the layer of proper skin (dermis). The epidermis carries a lot of bacteria, which would be risky for future cultures. Then, I placed skin proper on the bottom of a Petri dish, letting it dry and attach to the surface. After a few minutes, I had to pour, slowly and carefully, a little bit of the DMEM medium over it. The movement of the medium had to be limited because the biopsy is easily peeled off. After a few days, the migration of fibroblasts from the biopsy should be visible. It was hard to imagine that, but A. said it looked fantastic. (24 October 2016) (Fig. 5.4)

The two moments of the separation of my body elements were significant for my liminal projects and my assimilation to the laboratory network. From an emotional point of view, I started laboratory trajectories of

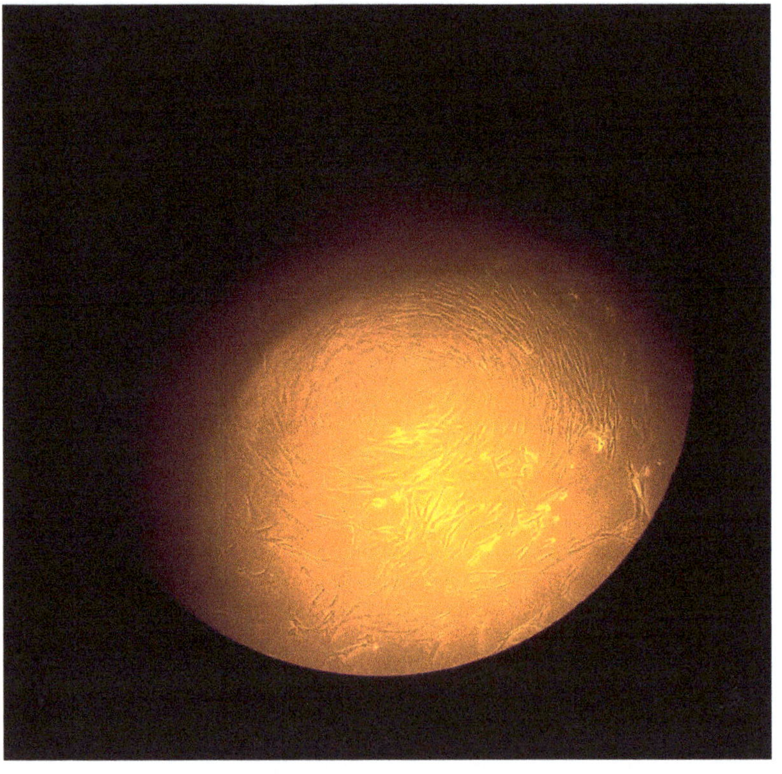

**Fig. 5.4** The migration of fibroblasts from the skin biopsy. Microscope picture by Karolina Żyniewicz

separate liminal beings, which obtained autonomy and the capacity to live in the technological bodies of incubators. My role was to take care of them, decide about their life span, and determine the position of their trajectories in those of mine. This also involved establishing relations with new actors of the biological network which were genetically related to me. To define my isolated cells as human or non-human liminal beings was challenging and confusing.

Day by day, I was building my relations with the cells, learning the laboratory language, and trying to blend in with the comings and goings of experimental research. If I heard a new word, I immediately asked what it meant or looked it up on the internet. The regularity of my tasks helped me to improve my skills. The biologists working with me tried not to infantilize their language when explaining particular activities to me. I often made them aware that their talk was hard to understand.

> J. explains things using specific scientific terminology and often forgets that he is dealing with someone for whom it is not daily bread. However, this is an effective method for quickly introducing me to issues, one far better than infantilization. (25 May 2016)

Besides capturing transmatterive moments on the biological level, I realized that I needed to understand and master the scientific language and try to transmatter it to be understandable in the two other contexts: art and academia.

While working with my cells, I visited various laboratories to conduct different experiments. To develop *The Last Supper* project, I had to carry out experiments in a laboratory researching mitochondrial gene expression in yeast, whereas the *safe suicide* work was mainly based in mitochondrial diseases laboratory. Consequently, over time, I collaborated with multiple research teams, learning their different customs. I only spent enough time to feel somehow at home in three laboratories. Nevertheless, this sense of belonging and at-homeness became highly important to me. It is an integral aspect of the liminal position that a liminal being sometimes feels that it would be much better to be located in a more structured and defined field with which they are co-vibrating.

> Teamwork is perhaps the most beautiful part of a scientist's practice. I envy them very much, which is why I try so hard to feel part of the team. However, it is impossible; my thinking goes in a different direction. People

> are beginning to get used to my presence, but they do not talk to me the way they speak to each other. Still, sometimes they forget to simplify some terms and apologize when they realize I am not following. (30 May 2016)

In a seemingly contradictory development, the sense of being a stranger and vanishing in the group appeared practically at the same time. At a certain point, I learned to be in the lab on my terms, without desperately striving to be 'like the biologists.' This position, however, was not understood in the same way by everybody:

> It is not easy to pinpoint my attitude. My saving grace is that I am doing my PhD; it rationalizes my lab activities in the eyes of the scientists. Developing an artistic project alone would perhaps be more difficult to swallow. It is noticeable that the scientists have their doubts about whether what I do is art. Some are uncertain but helpful, others are sceptical and reserved, and yet others are neutral. I do my best not to disturb the natural rhythm of work, trying to blend in with the background instead. Sometimes, however, I feel strange, perhaps not like an intruder, but more like an alien from another planet. (31 May 2016)

It was exciting to realize how fascinated I was by the research work and how badly I wanted to mimic what the researchers were doing. It was a challenge to my liminal position, and the cognitive theatre almost sucked me in. However, it was apparent that, having no formal biological education, I would never become a genuine part of the laboratory team.

Working with several laboratory teams is undeniably even more challenging than working with only one. Being an ordinary actor can be exhausting when navigating different laboratory networks, especially when a liminal researcher wants to be well versed in the rules established by each network. It is also easy to become fascinated by science and wish to learn as much as possible. I attended scientific lectures, lab meetings, and conferences. There is nothing wrong with such an engagement, as long as one knows what one really needs to implement one's project, and what is only an exciting appendix to it. One tends to be powerfully overwhelmed, especially at the beginning of one's work. The experience involves acquiring new knowledge, meeting new people, and explaining the methodology of liminal practices to them. Repeating the same things over and over again every day can be tiresome. For me, the work became calmer with time, and I felt more secure and dared to refuse new lecture invitations. A moment came when I started to feel extremely weary explaining how my work differed from biologists' work to whoever I met at one or another laboratory. I could not complain that people were not interested in my liminal activity, but what I would really have preferred was to do my job being indistinguishable from other students at the laboratory. Some of the biologists even expressed their wish to contribute to my work:

> Interestingly, while doing their work, some people have started paying attention to what may potentially be of interest to me. Of course, they still believe that I am only interested in visual aspects, but it is nice that they want to share something with me. Perhaps I see something in their work that they stopped noticing over time. (23 June 2016)

People who closely collaborated with me began to understand my approach to art—and even gradually came to share it. This could be seen as my educational success. Besides offering me what they thought might be visually exciting pieces, they started developing ideas for new projects. Many of them seemed worth considering to me. The researchers knew which ideas were doable and which were purely abstract. At some point, my consultations extended beyond working on my projects to address texts I had written as part of my studies:

> Because whatever I write now requires touching upon biological knowledge, the biologists have come to correct my texts. It is funny because they are, as a rule, extremely meticulous, if not pedantic, about syntax and grammar. I have coined an expression to convey it: they correct my texts molecularly. (14 August 2016)

Giving these academic texts to the biologists for revision worked like inviting them officially to join my academic trajectory. I did so because I was afraid, I would otherwise misrepresent things, I wanted to be honest with them, and I appreciated the knowledge they shared with me. I viewed this as a matter of trust, an expression of respect, and evidence of my profound gratitude for their help and fruitful co-vibration. For them, it was an opportunity to have a glimpse at a humanistic/artistic mode of writing, which not infrequently addresses their work and often casts researchers as its main characters.

Collaboration with a liminal human being requires a special effort from laboratory teams. Biologists tend to specialize in narrow fields, begin their research by posing specific questions, and follow pre-defined procedures to answer them. A liminal practitioner comes with quite different queries related to their initial imagination. To answer them, biologists often must step out of their tracks. Some of them are happy to abandon their routines and embrace it as a cognitive adventure; others are annoyed and resent it as additional work to do. That is why when developing a liminal project, it is essential to find a laboratory dedicated to research into a related subject.

> Time and again, something goes wrong. I am at the stage of preparing specimens as a basis for microscope pictures. The problem is that the scientists cannot really help me to interpret these pictures. They have never needed to do experiments quite like mine, so they do not know what results to expect. Nobody has yet had the idea of cultivating adherent and non-adherent cells together (my B lymphocytes and HeLa cells). It is hard to fix non-adherent cells on the surface of specimen glass. Only polylysine can help with that. Fixed B lymphocytes are hard to differentiate from HeLa cells. It is also tricky to establish whether the result comes from poison administration or whether it's been caused by culturing stress in transport from one building to another. (15 July 2016) (Fig. 5.5)

Most liminal practitioners, including myself, feel like they are parasites in the laboratory—actors that disturb the established order. This feeling is partly related to what Latour calls 'studying up' and its effects, which he has observed in the attempts of sociology to investigate the life sciences, a discipline enjoying a superior social status (Latour 2005: 98–9). Similarly, art, the humanities, and liminal practices derived from them have a lower social rank and are considered less consequential and less rigid than the life sciences. That is why liminal researchers who enter the scientific field feel deeply grateful for a chance to obtain something from epistemic giants.

Finding a laboratory that is perfectly aligned with one's project does not guarantee that things will go perfectly well. Any actant of the laboratory network (living or non-living) can cause trouble. This tends to be more stressful to liminal practitioners, like myself, than to experienced biologists. They know which problems are minor, predictable, and easily solvable, and which are serious. This difference is neither obvious nor readily recognizable to outsiders. Laboratory devices and reagents are costly, so big problems also mean steep costs. Many times, I was scared like a kid who had messed up:

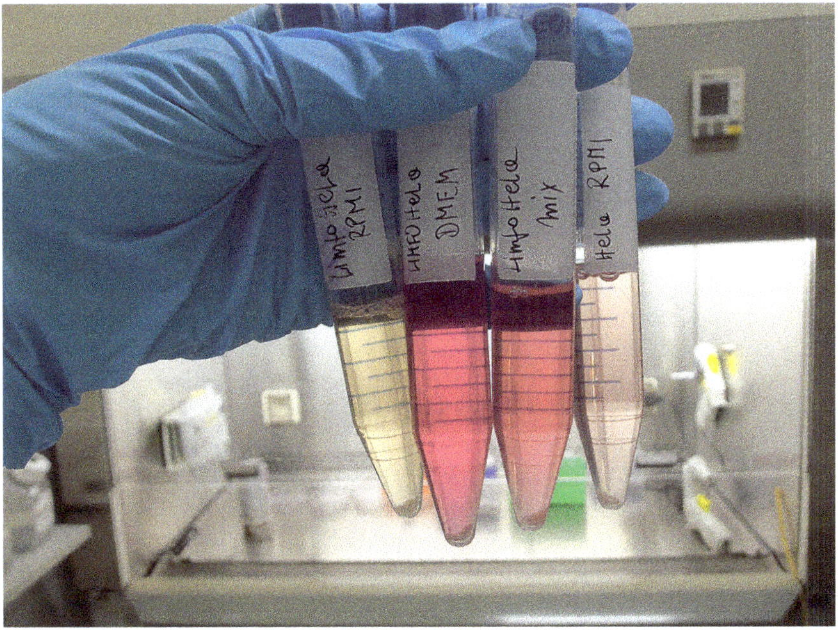


**Fig. 5.5** Joint cultivation of my B lymphocytes and HeLa cells. Photo: Karolina Żyniewicz

I've begun to get pictures of better quality, but problems with interpreting them continue. I need to consult many people. It hurts to have things apparently always beyond my control. In a couple of months of work, I've got somewhat unexpected results. Scientists have been trying to comfort me by saying that to have a total control of experiments is their great unfulfilled dream. Cells behave differently in every experiment. The only thing I can do is learn how to interpret the facts better. During the last microscope session, I lost a considerable amount of cells. I had to throw away one of the Petri dishes not to use up any more medium to keep the cells alive. I flooded them with hypochlorite, which works like hydrochloric acid and destroys living tissues. I could not work with each specimen, so they got wasted. It was all because of my poor organization of work. I need to work on that. Whenever something is wrong in the sterile room, I am afraid that I have caused trouble. An example of that was the incident with the pipettor. I realized that the pipettor in the laminar flow chamber number 1, where I was working, did not work anymore. It was connected to the charger but never actually charged. It was holiday time, so not many people worked in the room, only

> myself and E. I asked her what could happen to a pipettor that was uncharged all the time. She explained that the charger worked only when the laminar flow cabinet was running. Shortly afterwards, M. discovered that the pipettor did not work because the medium had flooded the filter. She said that I should have reported it earlier. I hadn't drenched the filter, but I felt guilty anyway. In the end, it was not a big deal because filters are replaceable. My independence hurt me in this situation. I do my best to avoid asking too many questions, but it is better to ask. (24 July 2016)

Establishing good collaboration is crucial for liminal projects. Members of laboratory teams are considerably independent in doing their parts of work. As already mentioned, they perform individual tasks and then bring their results together. Liminal practitioners cannot be as independent and need scientific supervision. Liminal researchers must do their best to master biological terminology and methodologies, but the interpretation of results is usually far too complicated. However, it is vital to collaborate instead of simply requesting biologists to do a given task.

The feeling of dependence may be difficult to cope with for people hailing from the art domain, where improvisation and creative freedom are crucial factors and values. These aspects are rather severely limited in the laboratory, and being ashamed to ask questions is frequently the reason for improvising. Sometimes, the 'out of the box' attitude and activities are fruitful, but more often than not, they cause real trouble:

> Today, I spent much more time in the laboratory than I had planned because of a problem with trypsin thawing. I needed to do a passage of my cells (fibroblasts). It is still hard for me to decide when the cells are ready. A. was cleaning the isotope room, so I had to walk back and forth to ask her specific questions. My cell culture is increasingly more prominent, so we decided to freeze a part of it next week. For today, we decided that I'd do a passage. I had to count the cells, but I forgot to take my lab notebook, so I could not remember the referential number. I also could not find a sheet with instructions, which was usually available in the room. I spent around 40 minutes counting the cells in a multi-well Petri dish and a culture flask. It was pointless because the number of cells in the Petri dish did not correspond to the number I found in the flask. (30 November 2016) (Fig. 5.6)

There were many more occasions, like the one described above, when the work went to waste. It was not only my experience as a liminal practitioner; all biologists are familiar with it. Obtaining a good result is a

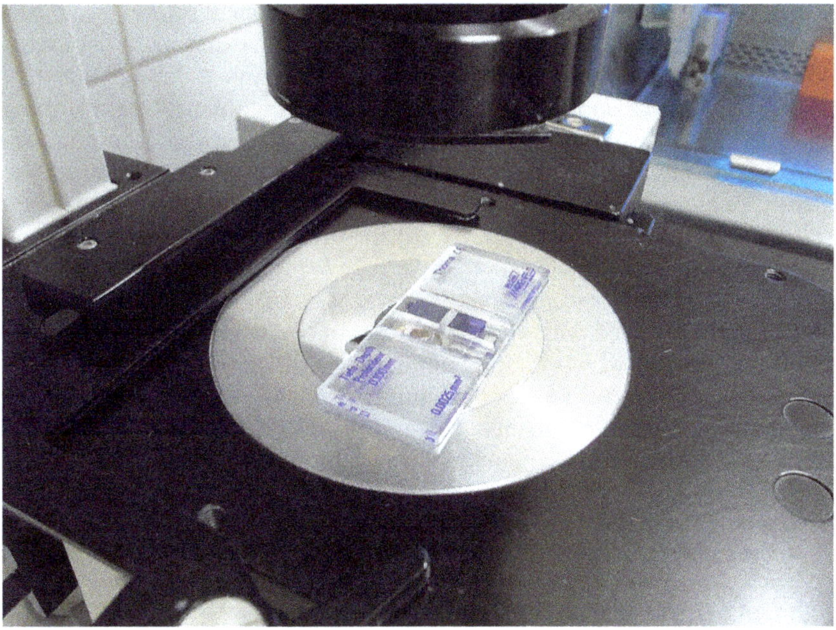

**Fig. 5.6** Cell counting in a Thoma chamber; cells stained with Tripan blue. Photo: Karolina Żyniewicz

moment of excitement and happiness, which often takes many months of work. Usually, the experiment must be repeated to confirm the result, but the effort makes sense. These small successes are as crucial to biologists and as they are to liminal practitioners:

> The moment of membrane development was inspiring. We were watching the picture appear. When we started to see white lines, we could be sure we had got our protein. We saw five lines. J. checked if they were on the proper level. They were, so it was evident we had five yeast strains containing our insert. It was a good reason to celebrate because we could start fermentation. (29 March 2018)

Scientists usually perceive failures in liminal practice as less consequential than those in their research. It is probably because they find it difficult to imagine what can and cannot be accepted in an artistic project. Additionally, the failures in liminal projects are as a rule more interesting visually than correct results. The creative component makes biologists think that such projects are done just for fun. I have met many biologists who had artistic hobbies and spent their leisure time painting, drawing, photographing, or composing music. At a certain moment, I was planning to organize an exhibition of their work, so I interviewed a few of such researchers only to find that some of them did not even try to share their artistic pursuits with their colleagues. They did not believe that it was a serious topic to talk about. They were surprised at my intent to focus on this part of their private lives.

Another surprising thing for them was that I did my artistic practice, which was associated with their free time, in the laboratory, which was a place of their serious work. Sometimes, when collaborating with a liminal practitioner, biologists are ready to improvise, which would be unacceptable in their scientific pursuits. The presumption that liminal practice is not a serious matter gives them a kind of freedom:

> Whenever I plan to multitask, my cells have different plans. Today, I came to the Institute at 9 am to prepare stuff for my appointment at the microbial laboratory around 11 am. My cells nastily did not want to detach from the bottom. I was waiting for more than 20 minutes to have them partly detached. I decided to centrifuge what I got without counting. The pellet was tiny, so I supposed that the number of cells was negligible. I was waiting for R. to come, afraid to see my cells under the microscope. We did it together when he'd arrived. My suspicions were correct. I had very few cells, too few to prepare the experiment correctly. We decided to place all cells on the same Petri dish; I knew that it was risky as it increased the chance of infection. However, no better solution offered itself. Our next problem was that there were no specimen glasses of the right size. Surprisingly to me, R. started to break specimen glasses that we had found in a random drawer on the edge of the bench. Seeing that, I was sure we would not get any good results as the specimen glasses placed on the same petri dish overlapped, especially while pouring the medium. I was trying to separate them using pipette tips, but then I needed to carry the petri dish to an upstairs incubator, so the movement made the glasses lie on top of each other again. In a sense, I should have liked this work style, so-called artistic chaos, and improvising. Still, I had spent enough time in biological laboratories to prefer an ordered working style based on procedures. We decided with R. that I would come tomorrow to check if the cells were ready for the experiment, but deep inside me, I was sure it was pointless. (20 February 2017)

This behaviour was strange, even for an artistic methodology. The scientists working with me were usually surprised to hear that there was something like an art methodology in the first place. Their astonishment was related to their notion of art as something not serious. With time, the biologists, having an opportunity to observe my work for an extended period, understood what I meant by an artistic methodology. In my case, this methodology was primarily based on their own work methodology.

As an artist, I graduated from the Academy of Fine Arts and specialized in sculpture and environmental art forms. It was always evident to me that scientific practice relied on the same methods that I used in my sculpture studio. Sculpting began from a general idea, sketches, articles, materials, and information about possible mediums. Then, I had to survey this assemblage to choose what I needed as a roadmap for developing the final form. These phases of accumulation and selection alternated in the creative process, like they did in scientific research. It was a matter of deciding that the thing was completed and I was ready to show it to the world. Biologists also have to decide when to stop experimenting and take their time to write a research paper. Then, of course, in art and science, the cycle continues, and the creative/conceptual process is restarted, with the thing previously defined as completed subject to further transformations and reconfigurations.

Another aspect of the artistic methodology is that biologists are not easily convinced that the ethics of liminal practice requires doing laboratory experiments properly. Admittedly, a liminal practitioner's goals may differ from those pursued by scientists, but it is a mistake to take failures in the production of liminal projects too lightly:

> Today is an example of a typical Friday the thirteenth. I had changed the medium in my cells last Friday. Today I saw strange objects around them. A. had a look at them and said that it could be a sign of infection, and it would be better to remove this petri dish. I was concerned because I knew I had only 3 test tubes with cells left in the freezer. A. was trying to comfort me by joking that I still had the whole skin of cells covering my body. It was not helpful. Additionally, it turned out that I had been using a medium without antibiotics all the time. After opening a new medium bottle, I had not added them because nobody had told me to do that. It was such an obvious thing for everyone else that it was not worth mentioning. Nobody was surprised by that, but I was sad that I had unintentionally lost my cells. (13 January 2017)

The incident above shows that, though essential, incorporating a liminal practitioner into a laboratory team is problematic. After some initial training, the liminal researcher becomes invisible and does their work like everybody else. That is why biologists fail to share certain information with the person, assuming that this is common knowledge. It can be a matter of tacit knowledge already imprinted in the body so firmly that people do not even think about it (Polanyi 1966: 15–6). Their actions are automatic. The failure to share information is not only a problem faced by liminal practitioners but also an experience of biologists who come to a new laboratory and join a new team, or even biologists who have worked together for a long time but try to use new methods described in research papers. Many details are not included in officially published protocols; they are the know-how of each laboratory. One of my observations in laboratories was that everybody kept a lab notebook that had to be constantly available to the team. People would not take their notebooks home in order to make it possible to their colleagues to find the information they might need. Often, when scientists moved to another lab they would leave their notebooks at their old workplace. I saw notebooks neatly kept, for others to quickly locate the bit of information they needed, while other kinds of written documentation were messy and difficult to consult for anybody else. I also kept a separate lab notebook for each of my liminal projects and stored them in the laboratory. Nobody was likely to use them in my absence, but occasionally I was asked how I had done one or another experiment, such as how I had prepared polylysine glass coating.

Misunderstandings and mistakes linked to tacit knowledge characteristically cause many problems before being discovered:

> I had a problem detaching cells again, so I asked A. if that might be a problem with trypsin. She went to the cultivating room with me, and we found the cause. I had been repeating the same mistake for some time. After removing the medium, I should have washed my cells with PBS before I poured trypsin in. The medium was blocking trypsin action. That was why it was so hard to detach the cells. When I did, what A. suggested, the cells got separated in fewer than 10 minutes. (21 February 2017) (Fig. 5.7)

The participation of scientists in the production of liminal projects is rather evident, especially when they help in situations such as those related; meanwhile, the involvement of liminal practitioners in research done by laboratory teams is relatively rare. However, it does not mean that liminal

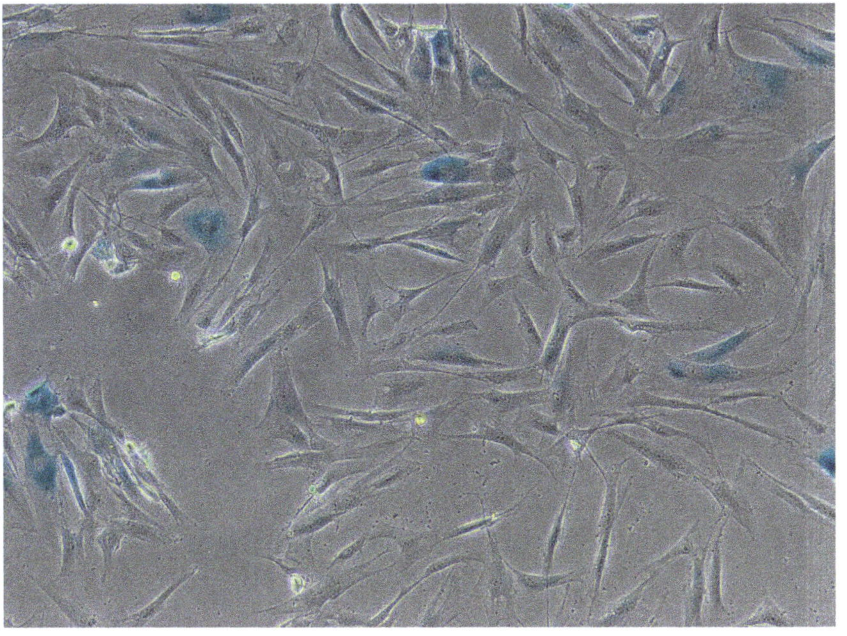

**Fig. 5.7** Very dense fibroblast culture requiring passaging; cells stained with B-gal. Microscope picture by Karolina Żyniewicz

projects do not have any cognitive value. Sometimes, the execution of liminal projects produces results of interest to biologists, but they often cannot use these data. Scientists cannot give their full attention to random discoveries because, busy doing their regular research, they rarely have time for such distractions. Nor can they quote a liminal practitioner in their publications as they are expected to cite solely verified and repeated results whereas repetition would entail abandoning the liminal position. However, random discoveries in liminal projects can inspire scientists by breaking their routines:

> The cells which I thawed on Friday did not look well today. The good thing is that a few of them are attached to the bottom, and the medium is not cloudy anymore. The question that popped up was whether I should add antibiotics to the medium I had been using so far or whether it was better to prepare a fresh medium bottle. I was sure the old bottle was without antibiotics, so it was risky for further cultivating. On the other hand, I did

> not want to waste the old medium. People from the lab convinced me in the end that it was better to waste some medium than to do the whole experiment all over again. I decided to work with the old medium on a spared portion of cells, but I would open a new medium bottle for the main culturing. We realized that all technical problems I had recently had could cause interesting changes in cell morphology. The cells had to fight hard to survive, having limited nutrients and not attaching to the bottom the way they should have. Chances were that they established unusual relations with each other. It was a shame we could not check it properly. The experiment would have to be deliberately repeated. (16 January 2017)

In my collaboration with biologists as a liminal practitioner, I was often bothered by the lack of adequate financial resources. Even if I managed to secure some funding, the amount was ridiculous as compared with the cost of laboratory work. If it had not been for the biologists sharing resources from their grants with me, I would not have been able to do much. Fortunately, I collaborated with the biologists who fully understood struggles involved in obtaining funding and supported me whenever my attempts failed. I was lucky enough to only meet scientists who were open to and curious about my practice. Many procedures necessary to further my project succeeded due to the snowball effect. Some scientists brought me in touch with their colleagues who could help me one way or another. The network was becoming increasingly more significant. I also felt that I had to give them something in return. At a certain stage in my work, I became confused in my best efforts to make the scientists grasp the 'meaning' of my work. I wanted them to think that what they were helping me with was a valuable endeavour. From my perspective, the initial 'transaction' of selling the project value was not enough. I felt I needed to keep the biologists believing that they had done a reasonable deal. Some of them told me they saw the value of what I was doing and that I should not feel bad about not giving them anything back. The very sharing of experiences and knowledge between us was meaningful to many of them.

Liminal collaboration—co-vibrating, when one of the parties involved is 'an outsider' from the liminal zone—abounds with excellent exchanges, but also with misunderstandings. In my experience, such slips were usually caused by communication difficulties ensuing both from my ignorance or miscomprehension and from language differences. As a result, *The Last Supper* project proved much more complicated than *safe suicide*.

Communication glitches may have been a nuisance, but they were also interesting. Conversations with the biologists often appeared more valuable to me than the project I was carrying out with them. I was always process-oriented.

A liminal practitioner's immersion in the laboratory environment is essential for transmattering processes. The immersion is founded on accepting the structure of the epistemic field and, at the same time, on remaining in the liminal position, with all its frictions and setbacks. Clarifying to scientists what liminal practices are and what they are all about is another prerequisite of successful collaboration. In my case, it was expedient to explain that I was an artist and a researcher, affiliated with the university and carrying out an autoethnographic project. The connection between these roles can be far from obvious to scientists who work with different methodologies and in different styles. Being transparent and trustful is the only way to achieving the liminal goals.

Before entering a laboratory, one tends to imagine what it is like, and these imaginations are bound to differ from the reality one comes to experience. One usually hatches a scenario inspired by sci-fi films, in which laboratories are furnished with state-of-the-art devices and operate in an almost magical way. Perfection is another common preconception, as the cinema produces an illusion of everything being planned and achieved on time, with biologists in control of all the aspects of experiments. Briefly, one forms visions that are idealistic; at least, mine was like that. I thought I would be able to ask biologists about everything, and they would know all the answers. I wanted to trust them. Fortunately, this idealistic approach disappeared soon enough. I understood that abiding by the protocols did not mean everything would go to plan. Working with living matter is never entirely predictable, which means that the transmattering of biological matter cannot be perfectly planned. This unpredictability is a facet of the plasticity of matter, which I describe in the next chapter. The fact that an experiment is repeatable does not mean that everyone will repeat it easily and unproblematically. The risk of mistakes is high throughout the experimental phases and activities. Biologists taught me that working in the laboratory required a lot of patience and tolerance of failure.

> I learn slowly, day by day, coming to the Institute from another part of the city, sometimes just to do a task that only takes 15 minutes. Often, I cannot even do my job because my cells seem to have other plans. It is a big lesson of discipline and humility. When doing sterile work, you must be very careful

> all the time. Also, everybody can just have a bad day when things simply go wrong. For instance, I confused the flasks with immortalized and normal cells in one of the passages. Consequently, I started to count immortalized cells, surprised that their number was increasing instead of decreasing. Fortunately, A. noticed my mistake, but a lot of my work went to waste anyway. (19 April 2016)

Sometimes one mistake was not enough, and I only learned my lesson after I had made the same mistake many times over. It is hard to say which laboratory agents, living or non-living, caused me more trouble. Strangely enough, if an experiment had to be completed to meet a deadline, non-human liminal beings tended to behave in unexpected ways. Another problem which I faced was that often a mistake is not immediately apparent. It usually took some time before I realized why something had gone wrong. An investigation was needed, searching for an actant that causes trouble. Mistakes were rarely identical, which made such investigations even more complicated. Sometimes, when trying to identify the reason for one error, I ended up making several new mistakes:

> My cells refuse to collaborate. We observe things that are not entirely understandable to anybody. It is always surprising that something is surprising to scientists who have worked with certain biomatter for many years. The portion of cells that thawed the day before yesterday is not visible under the microscope. It looks like I only froze the mixture for freezing, without cells. A. is on holiday, so I decided to show my cells to E. She was tentative in giving her opinion because she usually works with non-adherent cells. In the end, she suggested that I remove the suspect petri dish. I thawed another portion of cells, but that one also looked strange. The cells were not able to attach to the bottom of the petri dish. I called A., but she could not tell what was going on from a distance. I was devastated, knowing that I had only five cell portions more in the freezer. I could not lose any more cells. That would mean starting everything again from scratch, including the biopsy. I felt helpless because nobody could say: Do not worry, I had a similar situation. Failures in a creative process can mean success but not really in artistic production based on science. I decided to leave the cells on a petri dish until tomorrow, feeling that nothing would go well this week. (5 January 2017)

Even if one is a specialist in a given field of knowledge, it does not mean that everything one does will go smoothly. Sometimes, the cause of failure is so trivial that nobody would even consider it. Protocols include several steps to follow, and identifying where a mistake occurred often means going through all of them all over again:

> I came to suspect that my problems with my fibroblast culture could have something to do with the material Petri dishes were made of. I've recently got a few boxes of Petri dishes from the producer to test them in my work. I've tried to run a cultivation on them two times and failed both times. I noticed the liquid of the medium behaved in a strange way. It looked like the bottom of the Petri dish had a layer of fat on it. I decided to transfer my cultivating to Petri dishes that were regularly used in the laboratory. My intuition was correct. The medium covered the bottom of those Petri dishes quite well. I asked A. to have a look at the cells a few hours later and tell me if my cells had started to attach to the bottom. After some time, she e-mailed to tell me that the culture was working well. (9 January 2017)

It turned out that the company produced two types of Petri dishes: one for the cultivation of adherent cells and the other for non-adherent cells. I did not realize the difference and cultivated my adherent fibroblasts on the Petri dishes for non-adherent cells. Communication with the company confirmed this finding.

Mishaps and blunders, such as my Petri-dish bungle, are to be expected when handling all other laboratory devices: microscopes, Polymerase Chain Reaction (PCR) machines, pipettes, scales, etc. Every user can deliberately or inadvertently change one or another parameter in their settings and, in this way, make life difficult for the following user as restoring the previous settings is usually time-consuming. Worse still, the person who made the changes may be unable to retrace their steps, which further complicates the resetting. As a liminal practitioner, I was always reluctant to try and reset a device for fear of breaking it:

> I realized that something was wrong with the microscope. The light adjustment knob was set to the maximum value. I did not know how to set it back to the usual position. I supposed that the lamp was broken. I looked up who was the last to use the microscope yesterday. By doing that, I wasted the time reserved for observing cells. I was hoping for somebody to turn up to ask them for help. Unfortunately, nobody was coming up. Additionally, the incubator started to beep. Finally, A. came into the sterile workroom when I was doing centrifuging. She tried to restart the microscope, but finding the previous settings was difficult even for her. I could barely see the lines of

> the Thoma chamber, so the counting of cells was very imprecise. (28 February 2017)

It is challenging to determine whether a result is wrong because of a device malfunction or because of an error in following the protocol. I was never sure if what I was seeing was what I was supposed to see. With time, I understood how subjective laboratory research was and how much it depended on scientists' individual perceptions. As Latour and Woolgar observe, "the elimination of alternative interpretations of scientific data and the rendering of these alternatives less plausible is a central characteristic of scientific activity" (1985: 36).

Because I had to economize as much as possible, I sometimes decided to perform quite risky operations that biologists would probably never do. Sometimes the risk paid off, but, more often than not, it proved irrational and not really productive:

> Because I had used up half of the medium bottle I'd been given by M., I decided to rescue the previous bottle by adding antibiotic and Fetal Bovine Serum (FBS). A. gave me a flask with FBS and two flasks with leftover antibiotics. I mixed that with the medium and gave this mixture to my cells, which were going to waste anyway. If everything were ok with them by Monday, it would mean that the venture had succeeded. I should not have boasted about this attempt to M., who had objected to saving the medium from the very beginning. She said that it was not really saving because FBS was considerably more expensive than the medium (which I did not know). (17 February 2017)

As already stated, I usually admitted to my mistakes with the face of a kid who had done something wrong. Biologists laughed at me then. Yet when I was going through a crisis and things were not working for me, I sometimes did not want to share it with anybody. It was not a matter of shame; it was being tired of failures:

> I could not leave the Petri dish unwatched for the few days that I was not going to turn up at the laboratory because of a seminar and a conference, so I decided to pour hypochlorite on the plates. It was not an easy decision at all, and it made me feel like a murderer. After a few minutes, the cells exploded, which made me feel even worse. I did not tell anybody in the lab what I had just done. (8 June 2017)

It was crucial for building the temporalities of my projects to learn that laboratory work was based on counting the time and waiting. The timer can be regarded as the most critical laboratory device. Protocols and time in fact determine the order of scientists' life. Often, a protocol dictates when biologists can eat, go to the bathroom, or go home. Everything has to be done unhurriedly and carefully. Rushing up often results in mistakes. Procedures can only be stopped at certain moments, for instance, when a prolonged incubation can be left overnight. I was fortunate in not having very short project deadlines because otherwise it would have been tough to align the project and laboratory timelines. Not all experiment results appear on time, and not all can be accepted.

Artefacts are a relevant aspect of 'outside the order' laboratory situations and the differences between the approaches of biologists and liminal practitioners. In my view, artefacts are abjectual transmattering agents. Before entering the laboratory space, I associated the term 'artefact' with archaeology and art and thought of it as something that humans have produced. Generally, artefacts appear to be less appreciated than artworks in art. In the life sciences, the appreciation of artefacts is even scanter, and they are considered problems, mistakes, and misunderstandings. Michael Lynch has classified artifacts as 'positive artifacts,' 'distortions,' and 'negative artifacts' (1985: 81–122). Positive artifacts can be called harmless artifacts. As Lynch has defined them, they can be detected with relative ease and accompany a valuable result. Scientists are accustomed to positive artefacts whose occurrence results from the specific working of measurement and visualizing devices (1985: 81). Lynch's examples of these 'friendly' artefacts include: 'staining artifacts,' 'knife marks' (linear streaks or holes), 'folds,' 'burn marks,' 'strains,' 'foreign matter' (1985: 92–3). All of them are mechanical and pretty evident. A bigger problem, as Lynch observed, is with 'distortions,' they are more challenging to detect, because "they did not appear as external intrusions upon the 'material'" (1985: 98). Lynch's list of 'distortions' comprises: 'stretching,' 'optical aberrations,' 'fused membranes,' 'broken membranes,' and 'blurriness' (1985: 99). Lynch considers 'negative artifacts' the most dangerous enemies of laboratory researchers. As he evocatively explains: "Negative artifacts are available as nothings or absences; work which fails to get the thing to appear" (1985: 119). This kind of artefact is sneaky and very misleading.

In terms of my approach, artefacts result from a rebellion of non-living actants. I learned to recognize positive artefacts and distortions relatively fast. 'Negative artifacts' were too complicated for me to detect. They looked like credible data to me. As already mentioned, liminal research tends to be regarded by biologists as something less rigorous than their regular research. Many of them believe that artefacts can be exactly what liminal practitioners look for because they are visually appealing. Consequently, biologists are often surprised to find that an artist/researcher is not necessarily enthusiastic about artefacts and is not going to transmatter them as part of art production. Of course, artefacts as evidence of human imperfection can be attractive, but they cannot be included in liminal projects, at least not in my practice. One needs knowledge and experience to distinguish an artefact (mainly a 'negative' one) from an accurate result; this is one reason why liminal practitioners depend on scientists.

For a liminal practitioner like myself interpreting microscope images on their own is always a complicated thing to do. One constantly learns how to see and understand them, as without these skills they are only beautiful abstractions:

> W. helped me to do double staining: Hoechst and iodide. We had a problem identifying the results of staining under the microscope. We consulted a few people and concluded that changes in the nuclei were visible and that the cells might not be apoptotic yet, so the iodide did not work. Another possibility was that I had used too little iodide. The team had never done the procedure on healthy cells. We also considered administering Trypan blue, but we concluded that it destroyed cells so fast that it would radically shorten the available observation time. I decided to keep the pictures of the nuclei and the images done in transmitted light, in which the shape of the fibroblasts could be seen; watching those microscope images together with the biologists, I was learning how to make sense of what I saw. I could not be satisfied merely with a lovely photograph anymore. Beauty alone would not be complete. (8 February 2017)

Except for the interpretation of visible results, a massive part of laboratory work is not based on anything observable. The paradox of laboratory work is that complicated processes are hidden under ostensibly effortless actions, and this is hard to imagine. For me, doing a liminal project based on genetics or synthetic biology often means repeating certain procedures without fully understanding their implications. Therefore, scientific collaborators are navigators and guides. With non-visual results, recognizing artefacts is practically impossible for an artist/researcher:

> I was blocking a membrane in milk, and J. tried to explain how the process of blocking worked. I understood that proteins were located all over the membrane. We needed antibodies to attach to just a few of them, so the rest had to be blocked. I tried my best to imagine these relations between proteins and antibodies, but it was not easy. The last task for today was setting the primary antibody incubation. The result of that was another thing I had to imagine. (28 March 2018)

A typical example of 'positive artifact,' which I had met in my work many times, are air bubbles. Bubbles are as a rule beautiful visually, but they can ruin an immune specimen or electrophoresis. Bubbles can be mistaken for cells if the microscope is focused on the wrong layer. I collected artefacts for my research as a separate category. I thought of them as by-products and a kind of junk, which might be used in the future (or not).

> Bubbles are the greatest enemies of a biological lab worker. They can ruin even the most critical experiments, tiny, silent actants that they are. (28 March 2018) (Fig. 5.8)

Because not all biological research involves taking pictures, laboratory researchers may not even have good-quality microscopes at their disposal. They may also be not familiar with microscope photography and, consequently, not well versed in identifying artefacts. Entering a biological laboratory as an area where I could transmatter I-with, I was not prepared for obstacles such as artefacts. Technically speaking, they did prove that something had happened to my personal matter, but they could not be acknowledged as scientifically valid or valuable.

Having completed the second stage of my *safe suicide* project, I decided to stay on in one of the laboratories to do a regular internship. It was summertime, and there were several interns, mostly biology students, at the laboratory, so it was not easy for the biologists to find things for me to do that would be useful to them and, at the same time, not risky if I did something wrong. The point was that what I did was not some redundant research anymore, and my mistakes would be 'serious.' For my part, I was learning knowledge I did not really need because, as a liminal practitioner, I was not supposed to remain in this laboratory or in this field of biological research for long.

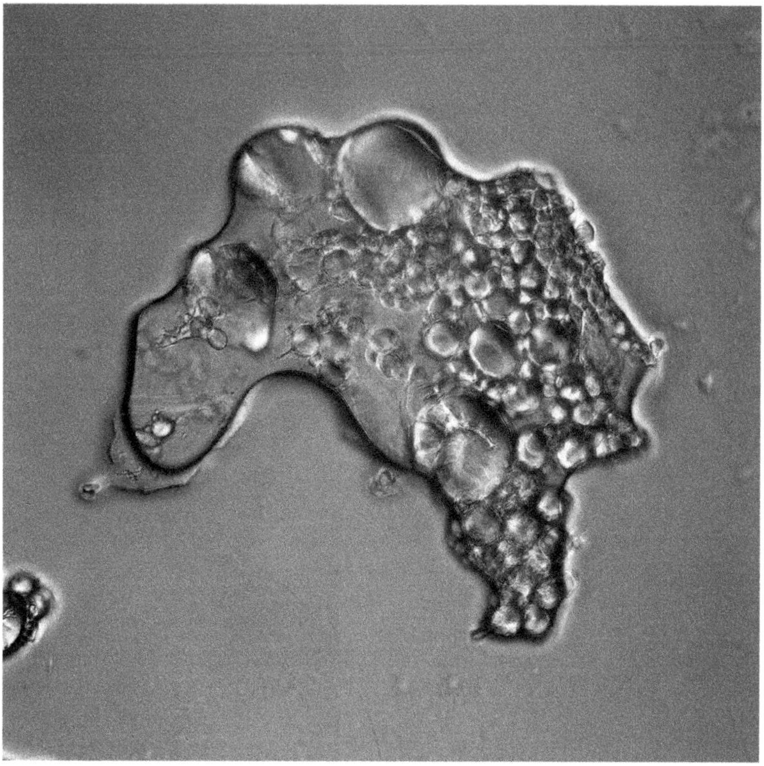

**Fig. 5.8** A bubble artefact; a confocal microscope picture. Laboratory of Confocal Microscopy, Faculty of Biology, University of Warsaw. The joint cultivation of my lymphoblasts (lymphocytes B immortalized with Epstein-Barr virus) and HeLa cells (Flipln) cultured in a mix of mediums (RPMI and DMEM), after the administration of hypochlorite; the image shows the drying out of the specimen with cells hidden under air bubbles; its pictorial appeal is of no scientific value. It is difficult to establish which of the cell changes have been caused by hypochlorite as drying out is a lethal factor in and of itself

She could not give me too complicated tasks because the samples were too important to risk losing them. That was a different situation than working with my cells. I felt that somehow, I was wasting my time. Biologists do what they do for a specific reason, to conduct their research. In my case, I was learning how to do things without a clear perspective of using this knowledge in the future. People in the laboratory were amazed that I

> decided to devote my time and energy to that. I was more trouble than help to them. (2 August 2017)

My intention to work for scientists was good, but it turned out that they did not expect that of me. Finding something for me to do and supervising what I was doing were additional chores for them. They asked me to take care of a self-immortalized muscle cell line, which had earlier been cultivated by another intern. I had to observe the cells, count them, and put the data into an Excel table. It was quite hard to continue an experiment started by someone else, but by that time I was quite skilful in cell culture.

> I can say that my work with this self-immortalized cell line made sense. The cells looked interesting, and I could learn cultivation without antibiotics. Observing these cells was not crucial for the laboratory, but they were curious about what would happen to them in a longer perspective. My other main internship activity was working with the Western blot method. This activity made less sense than doing the cell culture. Even if I became well versed in Western blot preparation, I would not have any opportunity to use those skills. My internship advisor treated me just like the other interns, which had good and bad things to it. The advantage was that I learned a lot. The downside was that I was forcing myself to know as much as biologists. I was ashamed when I did something wrong, and I was also concealing my liminal position. (3 August 2017)

The internship was one way of 'giving myself to science.' I had to put my transmattering intention aside and work on I-with for a moment. I worked in the service of somebody else's concept. When working with my own cells, I would also give other tissues of mine to my laboratory colleagues from time to time as so-called control specimens. They were cells, urine, and saliva. I also offered my skin biopsy to the doctor who had performed the biopsy to isolate fibroblasts for my project. She was grateful, but she did not contact me in the end. When my work on *safe suicide* was over, I decided to prepare a new line of my B lymphocytes to give the cells to my scientific collaborators as a backup for their future experiments. They would use them whenever they needed. A test tube with my cells remained frozen in each laboratory with which I collaborated. This second B lymphocyte line was intentionally stored. I can use these cells for my future projects, but the lab scientists are their primary owners. This is the way in which I wanted to repay them 'in nature' for their help.

> Together with M., we retook my blood to isolate and immortalize a new B lymphocyte line intended as a gift for science. M. said that she was not skilled enough in blood taking, so one attempt might not be enough. We found containers for blood collection; they were slightly dented, but we did not think it would matter in that situation. M. tried to take blood from either of my arms, obtaining tiny amounts of liquid each time. After a few attempts, she proposed me trying to take her blood to see whether it was the matter of her skills. I was experienced in working with blood, but not in taking blood. I preferred being a donor. In the end, we collected enough blood for cell isolation in a few containers. I was excited to go back to work in the laminar flow cabinet. I was relieved, realizing that I remembered the steps of the protocol. (7 June 2019)

Donating my cells to science was the natural final step of my *safe suicide* project, and I thought of it as a symbolic action. The biologists promised to let me know when my cells would be used in an experiment. The general idea behind this is to see how my cells, elements of I-with, participate in the production of knowledge without me being present. What will happen with the test tubes with my blood stored in many laboratories is basically beyond my control. They may be used in one way or another, but they may also be removed during cleaning, and I cannot do anything about it. These liminal actors, my cells, live separate lives of their own. They may be found in a freezer or a liquid nitrogen container and used by somebody when I am already dead.

## References

Berger, Peter L., Brigitte Berger, and Hansfried Kellner. 1974. *The Homeless Mind: Modernization and Consciousness*. New York: Random House.

Glaser, Barney G., and Anselm L. Strauss. 1974. *The Discovery of Grounded Theory: Strategies for Qualitative Research*. Chicago: Aldine.

Latour, Bruno. 2005. *Reassembling the Social. An Introduction to Actor-Network-Theory*. New York: Oxford University Press.

Latour, Bruno, and Steve Woolgar. 1985. *Laboratory Life: The Construction of Scientific Facts*. Princeton, NJ: Princeton University Press.

Lynch, Michael. 1985. *Art an Artifact in Laboratory Science. A Study of Shop Work and Shop Talk in Research Laboratory*. London, Melbourne, Henley: Routledge & Kegan Paul.

Polanyi, Michael. 1966. *The Tacit Dimension*. Garden City, NY: Doubleday & Company.

Riemann, Gerhard, and Fritz Schütze. 1992. Trajektoria jako podstawowa koncepcja teoretyczna w analizach cierpienia i bezwładnych procesów społecznych. *Kultura i Społeczeństwo* 36 (2): 89–109.

Zakrzewska-Manterys, Elżbieta. 1995. *Down i zespół wątpliwości. Studioum z socjologii cierpienia*. Warsaw: Semper.

CHAPTER 6

# *Safe Suicide*: Plasticity and Transmatteringness

*Cell colonies, or organs, are events which progressively unfold themselves. They must be studied like history.*

*Alexis Carrel*

Squier describes some forms of laboratory living matter, such as stem cells, organs for transplantation, and banked embryos, as liminal lives. The common point for all of them is that they remain 'in-between' two bodies, two forms of embodiment (Squier 2004: 4). They travel between two organisms, and they stop in the laboratory for a shorter or a longer time. The laboratory is also the locus of the liminal transition and its time regulator.

The liminal status of my cells corresponded to my social and cultural liminality, as depicted in the previous chapter. In my research, I did not work with any of the laboratory beings considered liminal by Squire. That would have been complicated because of their value for medicine and the bioethical regulations in place. Nevertheless, I did work with cell cultures that travelled between the human body (my own body) and the technical bodies of incubators and freezers. These cells were forever suspended in the laboratory liminal condition. 'Forever' means here the whole time of their being kept alive through *in vitro* procedures.

While the previous chapter described the initial ideas of my liminal projects based on the becoming liminal of I-with, this chapter aims to relate in more detail what was happening with my I-with while


K. Żyniewicz, *The Body as Matter in Art and Science*,
https://doi.org/10.1007/978-3-032-13665-7_6

implementing the *safe suicide* project. In *safe suicide*, transmattering was mainly understood as transforming one liminal being (one body) into many post-liminal beings existing in laboratory conditions. My fieldwork, on which I draw in this chapter, was based on working with cells derived from my own body—my B lymphocytes and fibroblasts—and cells coming from other bodies (muscle cells and HeLa cells). Like the preceding chapter, this chapter is also built around the field notes I collected during the implementation of *safe suicide*. In my argument, I focus on my relations with cell cultures.

To properly embed my analysis of my relations with cell lines and their transmatterive potential, I must recount how cells taken out of the body obtained an opportunity of living independently in laboratory conditions in the first place. Their liminal existence is a tremendous scientific achievement, one that opened a new chapter in biological and medical research. Cell culturing makes it possible to examine intra-body processes outside the body. It is a relatively recent accomplishment, dating back to the early 20th century. Before, it was necessary to kill an organism to see what had happened after administering a substance. This form of biological research is still carried out in some laboratories, but tissue culturing has helped to reduce the scale of this practice.

Working with cells outside the body would not be possible without the fundamental idea of plasticity:

> Plasticity is an idea of living matter that is also a practical approach to it: Substance can be radically altered without causing death. Plasticity is the ability of living things to go on living, synthesizing proteins, moving, reproducing, and so on despite catastrophic interference in their constitution, environment, or form. (…) Although living things can be manipulated, part of the particularity of biological plasticity is that biological matter may change or react to intervention in totally unexpected ways. (Landecker 2007: 10)

Hannah's Landecker definition of plasticity was coined after Catherine Malabou had proposed her own one, inspired by Hegel's notion as reworked by the neurosciences (2004). Malabou explains that:

> So, to be plastic means that everything that happens to you fashions you but at the same time does not destroy you. Plasticity is the way in which time shapes or fashions us, constitutes our subjectivity and at the same time allows for resistance. (2016: 99–108)

Having in mind the aesthetic origin of plasticity, I imagined that matter could be shaped like a sculpture by adding (in clay modelling) and removing (in stone/wood sculpting). The bioart collective SymbioticA has tended to depict their objects as living sculptures. This understanding resonated with me as I had a background in sculpture. I believe that, in many senses, liminal practices reclaim the aesthetic function of plasticity.

Elsewhere in her book, Landecker enumerates three facets of biological plasticity. One of them is related to the plastic quality of living matter. It is flexible, elastic, and soft, all of which supports my sculpting analogy. Another dimension of plasticity that Landecker identifies is linked to the unpredictability of living matter. This facet is responsible for the fact that laboratory work only produces an illusion of control. Laboratory liminal beings are only controllable to a certain extent, and this aspect of their existence is an essential actant of the laboratory network. The last important facet of biological plasticity listed by Landecker is a form of practice in contemporary biology (Landecker 2007: 10). Today, biology is comparable to sculpting in living matter. In terms of Kunst's depiction of art as a production sector, sculpturing, both in biology and bioart, is a mode of the contemporary production of life.

Malabou proceeded from being interested solely in the positive aspect of plasticity to considering its destructive and traumatic dimension. She has acknowledged that: "Plasticity is both the form of our lives, the mode of being of our bodies, and what destroys us" (2016: 108). It means that we simultaneously embody both facets of plasticity as life and death work together in living beings. In all kinds of bodies, billions of cells die every day, and billions of new ones appear in their place. The balance of gains and losses keeps the organisms alive. Isolating cells from a body enables us to see this balance and watch the processes of dying and emerging. In my research, I relied on Malabou's theory of plasticity as mediating temporality and the future (2016: 102). This mediation is the same inside the body and outside it, in *in vitro* conditions. However, the temporality of these two realities is different. The lifespans of cells are different. Emphatically, *in vivo* and *in vitro* are not identical, and they must be understood as different theatres.

Developing the idea of extracting some elements from the body to keep them alive in artificial conditions entailed learning about tissue behaviour and, more importantly, devising cultivation methods. The basic technical and philosophical question behind the venture concerned the subjectivity of cells and its connection to the subjectivity of the body: "Each of the elements…of our bodies lives, without doubt, a little for us, but they live above all for themselves" (Landecker 2007: 29). This idea was central to *safe suicide* project as I wanted to know how my subjectivity and the subjectivity of my cells were interrelated. The relationship between the two lay at the core of transmattering. The project was envisaged as transmattering temporality and future: the temporality of my cells and the future of my I-with. It was mostly tinkering with biological matter and time.

To describe my own experiences with cell culturing and immortalizing I have to draw a bit of the background for these laboratory activities. The first meaningful experimentation with plasticity by means of tissue cultivation was undertaken by Ross Harrison and Leo Loeb (1907–1910). Harrison pushed the technology forward by advancing the 'hanging-drop technique' of tissue culture. He worked with frog nerve fibres, which he cultured in a medium containing blood clots, agar, or salt solution (Landecker 2007: 40–1). Besides the culturing method that made it possible to observe the tissues over time, Harrison used *camera lucida* to document experimental processes. This instrument enabled biologists to see images of the specimen superimposed on the paper and then to draw them (Landecker 2007: 42). That represented a very complex and coherent achievement.

Another important advancement in the history of tissue culture was accomplished by Alexis Carrel and Montrose Burrows, who perfected Harrison's methods and were actually the first to use the term 'tissue culture' (1910–1912). They could run secondary cultures by passaging cells that were already being cultured, which extended the observation time. The 'hanging-drop technique' was no longer sufficient, so Carrel developed the first tissue culture flask, called a D-flask, a prototype of the modern cultivating bottles. Carrel also realized that cells needed to be cultured in sterile environments, although antibiotics were not yet available. Carrel received the Nobel Prize in Medicine in 1912 in recognition of his significant achievements (Lijing 2010).

The major breakthroughs in the development of tissue cultivating indicate that it was a constant struggle with time constraints and that concerted effort was invested in making the observation process as long as possible. The top achievement in prolonging the laboratory lifespan was marked by inventing the method of cell immortalization. In what appears a highly symbolic gesture, science used the notion of immortality to suggest that the greatest dream of humanity—the removal of limits to life—was made a reality in laboratory *in vitro* conditions.

The story of cellular immortality has simultaneously been a story of relations between cells and the body from which they have been derived. The term 'cellular immortality' was eventually coined between 1910 and 1914, preceded by a range of phrases, ranging from 'reactivation' and 'rejuvenation' to 'regeneration' and 'permanent life.' Having kept embryonic chick tissues alive for days on end, Carrel called death a 'contingent phenomenon' and argued that life and death were controllable (Landecker 2007: 74). The processualism of cell life called for a processual observation method. The laboratory conditions spawned a new perception and understanding of time, and Carrel relied on cell cinematography to posit that biological time was something else than clock or solar time (Landecker 2007: 87). Intriguingly, biologists considered death to be a technical failure, a kind of artefact, instead of viewing it as the natural limit of life, one of two facets of plasticity. This has strongly influenced the general understanding of 'nature.' When biologists started to style themselves as designers and creators of living things, the importance of nature became considerably decreased (Landecker 2007: 92).

The development of cell immortalization commenced by Carrel reached its ultimate turning point in 1951, when Henrietta Lacks entered the Johns Hopkins Hospital as a cervical cancer patient. Without her knowledge or consent, the biopsy taken from her body for medical purposes was taken to the tissue research laboratory run by George Gey to be used in his cervical cancer research. Shortly afterwards, the biopsy, placed on a Petri dish, grew instead of dying. Gey and his team ended up concluding that they had discovered an immortal cell line. Henrietta Lacks died shortly after this discovery, and Gey's team failed to inform her family that her cells had proven revolutionary for the development of science. The donor's name remained unknown for a long time. When her identity was acknowledged, the cell line was given a designation derived from the first letters of Henrietta Lacks' name: HeLa.

The story long dwindled in obscurity, without attracting anybody's interest. It eventually came into the spotlight courtesy of Rebecca Skloot's book *The Immortal Life of Henrietta Lacks* (2010). Skloot carried out a detailed investigation into the two stories: that of Henrietta and that of the HeLa line. She reached out to the Lacks family as if trying to restore their right to be involved in what was happening with the cells taken from the body of their wife, sister, and mother. Skloot stressed that HeLa cells had not come from nowhere and that the body they had been harvested from had been a body of an individual who had lived in a particular social and cultural context. Skloot also showcased the dynamic approach of the Lacks family to what was going on with HeLa cells. For them, especially for Henrietta's daughter Deborah, seeing her mother's cells meant as it were seeing her mother. Henrietta was, so to speak, still alive for the family, living in the tiny particles. The book eloquently raised important questions concerning the independence and the general status of cells.

Essentially, at a certain point the HeLa story, which at the beginning was a story of enormous success and significant progress in science, took an unexpected turn. With time, HeLa has come to be seen as trivial and easy to cultivate, even in poor conditions (Landecker 2000: 57). The ease with which HeLa cells can be cultured, and their immunity to infections, makes them a dangerous thing that can contaminate other cell cultures. Another problem is racially coloured. Henrietta Lacks was a black woman. In the studies on HeLa contamination, the marker responsible for race was found to be the evidence of contamination by HeLa. Of course, this discourse was strongly related to racism, with the 'black' cells depicted as 'contaminating' 'white' cells. This connotation may have been one reason why the identity of Henrietta was concealed for so long. HeLa cells are so ubiquitous in laboratories worldwide that their numbers are difficult to calculate, but it is estimated that the weight of all HeLa cells combined gigantically exceeds the weight of Henrietta. Today, it is also difficult to determine how many HeLa cells come from the first line, directly stemming from Lacks's biopsy, and how many have been modified for further experiments. Another disputed issue is whether the modified HeLa cells should still be associated with their donor. Henrietta's body is constantly being transmattered and reconfigured.

The story of Henrietta Lacks and HeLa cells was one of the points of departure for my *safe suicide* project. My plan was to become the matter of transmattering and to use my own body's biological plasticity.

The human body consists of 'many,' even if being an integrated whole. However, in *safe suicide*, the selected 'many' were extracted from my body to acquire a form of independence. As already indicated, the theatre of life was adapted to the theatre of the laboratory. I was a director and an actor at the same time in this theatre. The script of my project contained many plot twists that I decided to call transmatterive moments.

The isolation of post-liminal beings from the liminal human body was like the first act of my transmattering spectacle. In the first step, I had my blood taken and isolated certain cells—B lymphocytes—from it. It was like getting some multiplicities from the organism and augmenting the organism. In terms of Deleuze and Guattari's idea of the Body without Organs (BwO), which can inspire one of many possible understandings of *safe suicide*, disconnecting some organic pieces from the structure of the organism was a partial enactment of the concept, since "the body without organs is never limited to an organism" (2017: 14). An organism is only one of the many layers that make up the body without organs and, even more interestingly, 'the self' is another layer of this construct (2015: 191). The intention of my project was to operate between these two layers.

I had to select certain components of the BwO multiplicity. Not every blood ingredient was functional; it depended on the goals of the project and on the profile of the laboratory. For example, erythrocytes, which were a by-product of B lymphocyte isolation, had to be removed because they could not divide and interfered with the culture of the useful cells. A selected component of the blood became a colony of independent post-human liminal beings. The prefix post- indicates here that the cells left a human organism to claim the laboratory area of post-nature. The laboratory trajectory as a whole comprised an isolation chain that I believe can be suitably pictured as opening a Matryoshka doll: blood from the body, cells from the blood, mitochondria or nuclei from the cell, and, eventually, DNA from both. That was multi-level transmattering. It looked like each layer of BwO could be sub-stratified again and again.

The process of taking blood was not long. We had to clean the laboratory bench immediately after the procedure. The cells had to be isolated the same day. I centrifuged the blood in a special Falcon tube (Leucosep 1632: 88). I could then see a few of layers in different colours. It looked a bit like a jelly (Fig. 6.1). To collect B lymphocytes, I only had to take the uppermost layer. The rest of them went to the trash bin, together with the tube. I felt like I was wasting something important. The biologists did not feel that way. For them, this is what is done in biological research: the material is selected to get what one needs for the experiments at hand. Most of the biomass, redundant life, is wasted.

> After working with blood, the whole sterile workroom had to be carefully cleaned up with ethanol, and the biohazardous trash had to be removed. I did the cleaning for the first time, and the smell of ethanol was overwhelming. 2/3 of the collected cells were to be immortalized using the Epstein-Barr virus. I did not immortalize the remaining 1/3 of them to observe the

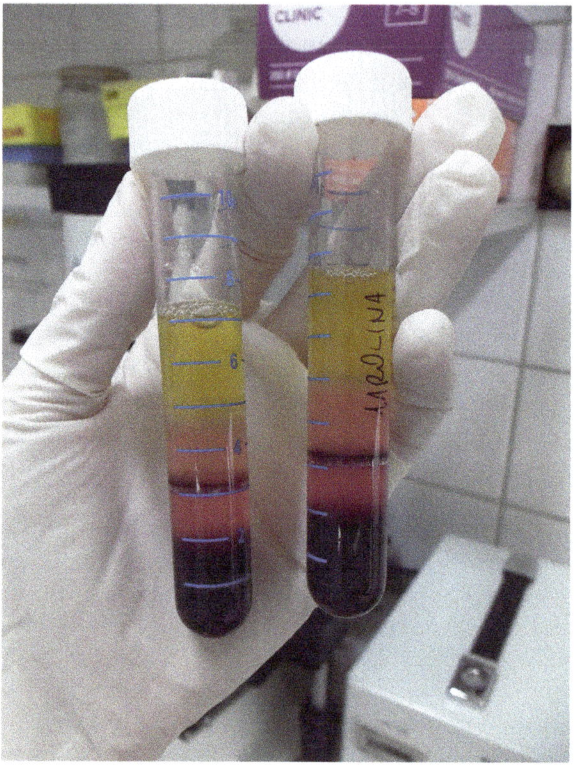


**Fig. 6.1** The separation of blood components. Photo: Karolina Żyniewicz

> natural process of their ageing and dying. Freshly isolated cells need time to adapt to the new conditions and start growing. (15 March 2016)

The very lopsided proportion between the amount of biomatter I obtained from the experiment and that which was thrown away makes it difficult to rationalize this kind of waste. I had to decide which cells would become my post-human liminal beings, knowing that the rest had to die. I was, so to speak, like Doctor Frankenstein constructing my own version of BwO. Erythrocytes were not going to be cultivated anyway since, having no nuclei, they would not divide. The biologists helped me to understand that it would be pointless to use expensive media in order to keep the cells useless in wet work alive. The idea of my liminal project presupposed the same criteria as all biological research, meaning that what was not useful for the research was to be eliminated, the way that layers of stone are removed by a sculptor to obtain the desired shape.

There was something ritualistic in working with my blood in this way. I was used to handling my menstrual blood, but I had never before worked with a blood sample taken in the laboratory. I had had my blood taken many times for medical reasons; on such occasions, however, I had left the sample for someone else to analyse. The *safe suicide* project made me feel that my blood was indeed mine because it had been derived from my body and because it had become my art medium. I was never scared of or disgusted by blood, but the fact that I had a tube full of something that kept my body alive was quite peculiar. In Kristeva's definition, blood outside the body is one example of the abject, and the abjectual quality was indeed captured in the blood taken and the isolated cells. The laboratory bench where I was to have my blood taken had to be prepared for the procedure. It was covered in metal foil and disinfected with ethanol (70%). The mandatory cleaning of the entire sterile workroom was even more significant; however, I performed the isolation mostly in one laminar flow hood. When I took the upper layer of the stratified blood and threw away the rest of the sample, my B lymphocytes stopped being abjectual. They became a part of the cast of my laboratory play.

The notion of death was pivotal to *safe suicide*. The project explored the meaning of death in *in vivo* and *in vitro* conditions. Time accounts for the fundamental difference between these two conditions. Life seems to be shorter and faster *in vitro* than *in vivo*. First and foremost, time is easier to measure and experience at the laboratory. It is not my goal to survey or contribute to the philosophical discourse on time; rather, my description of time is based on my embodied experience and field observations.

Before entering the laboratory, I had not known about the immortalization of cells. I was likewise unaware that there were two kinds of cellular death: apoptosis and necrosis. Apoptosis is depicted as suicidal cellular death, which was very important for my further work on the project. It is not that cells 'make a decision' about their self-destruction. The suicidal effect of apoptosis is based on a mechanism imprinted in the cellular genome. Internal cell impulses cause the dying process, and everything happens inside the cell.

For its part, necrosis takes place chaotically, caused by external factors. If apoptosis is referred to as suicide, necrosis can, by analogy, be called a homicide. Technically, the apoptotic mechanism can be initiated from the outside, for instance, by administering certain substances. However, if the intervention is too radical, necrosis ensues. Experiments conducted as part of my *safe suicide* project attempted to find the most accurate intervention. Metaphorically speaking, I committed homicides by forcing the cells to commit suicide.

The meaning of immortality in cell culture is somewhat different than in the macroscopic world. Cells that continue to divide interminably are considered immortal, regardless of the fact that the original cell is replaced by two new ones after each division. By that definition, most somatic cells (i.e. those not involved in gametogenesis) are not immortal (with stem cells being a rare, but very important, exception). Following isolation from the organism, typical human cells will undergo at most 50 divisions and then will stop proliferating to finally die. This phenomenon is known as the Hayflick limit (Hayflick 1965). The immortalization of a cell line means removing the Hayflick limit and enabling the cells to continue to divide indefinitely. In nature, this is what happens in cancer cells, which explains why the cells obtained from Henrietta Lack's tumour were immortal. Biologists have developed methods to immortalize cells in the laboratory; one of such methods consists in infecting cells with the Epstein-Barr virus.

The immortality aspect added a new dimension to my project; I could immortalize my cells (giving them a chance to divide endlessly) and also take immortality away from them (interrupting their lifespan). The immortalizing operation endows scientists with greater control of the cultured tissues. Their lifespan depends on the research schedule. Immortalized cells can die like regular cells in response to external factors, such as infection or limited nutrition. I found it intriguing that the prolonging of cells' capability to serve in research was named 'immortality.' This notion works in an interesting way when juxtaposed with other definitions of immortality circulating in culture. Immortalizing does not rule out the possibility of apoptosis, and cells can start the self-destruction process when the apoptotic mechanism is initiated.

Working with my immortalized cells was like performing on cellular temporality by adding or removing the future. Like the laboratory work in general, *safe suicide* was based on the perception and counting of time. Timers are the most critical laboratory devices as every protocol task requires an exactly determined amount of time. For this reason, it was essential for me to be conscious of time, to almost see it. How time is perceived in the laboratory depends on what devices (timers) show. In a way, devices capture time for the human brain. In the case of cell culture, the time is marked by the number of cell divisions. As recorded in the fieldwork note about the cell isolation, I decided to start two cell lines. One of them was immortalized, and the other was left unprocessed. I wanted to play with the option of being finite and infinite simultaneously. For each of the lines, the time was counted/measured differently.

> I've seen my cells today for the first time since the administration of the Epstein-Barr virus one week back. The virus has reprogramed the cells, overcoming the Hayflick limit. From the moment of this reprogramming on, cells become like cancerous cells. It is paradoxical that something that gives immortal life to an immortalized cell line causes the death of the whole organism. Waiting for the final effects of the virus administration, I observe the other cells and count them in the Thoma chamber after staining them with Trypan blue. I have a problem finding the Thoma chambers under the microscope, so A. helps me with that. Trypan blue poisons cells, so even counting is a way of killing. I have collected the test tubes with the cells prepared for measuring as a relic. It is pretty surprising to the scientists. (22 March 2016)

By plotting growth curves, I saw two different cellular timelines. The number of the immortalized cells was increasing so I had to passage them time and again to prevent them from overpopulating the Petri dish. Immortality potentially extended the experimentation time into eternity. Yet it would have been boring and pointless to keep the cells alive forever. It could have been a theme for another project. This was a potentially unlimited temporality and letting it happen was consuming a lot of energy and media. I needed closure on my terms, a closure that was not the end of the play. It was possible to stop only a portion of the cells, letting the others continue to divide. The liminal performance went through many acts before the laboratory trajectory had run its course.

As a rule, the research schedule regulates the longevity of the play acts, and cells are no longer needed when the experiments have been completed. Even though they have survived examination and can be kept alive, their act is over, and the time comes for the next one.

> My cells are not ready for a passage yet. Today, I've asked in the lab what happens with cells when a study is over. The answer came fast and with a bit of humour: 'They end up in the sink pipe, flooded in hypochlorite, which is

> similar to bleach. What else is there to do? Should they be sent off to retirement?' Nobody has time or resources to take care of the material which is not helpful in research anymore; the matter that is being used requires effort enough. Cells share the fate of bacteria, yeast, and other model organisms. Scientists assume a rational approach to living research matter. They wouldn't be able to do their job properly, if oversensitive to every bit of living matter participating in their research. Biology is simply based on killing. (21 March 2016)

As suggested in the historical overview of tissue cultivation, this technological advancement was supposed to reduce the scale of killing because it made killing animals for all experiments unnecessary. Killing an animal was a definite closure. However, moving toward molecular experimentation did not fundamentally put an end to killing; instead, it shifted killing to another biological/evolutionary organizational level. It is much easier to kill a cell than to kill an animal. Post-human liminal beings do not evoke such empathetic solid reactions.

Immortality is an example of a term that is easily misunderstood outside the laboratory. The general understanding of immortality in culture connotes living without end, or rather in the messianic timelessness of 'kairos,' which Agamben interprets as "neither perfect nor imperfect, it is not the past or the future and is their reversal" (2009: 93). Arriving at this point seems to be one of the humanity's greatest unfulfilled dreams. For believers, the power to remove death lies with god alone. God can also grant one a second, eternal life after one has passed the border of extinction. In cellular immortality, humans can influence the lifespan of cell cultures and, therefore, humans in laboratory conditions can be said to be like gods. This reasoning usually informs social objections to biotechnological experiments. People, especially believers, do not want to have a human being elevated to the position of a god. They tend to confound the religious and the scientific definitions of immortality. Continuing the entrenched confusion of the *in vivo* and *in vitro* orders, cells can be considered 'victims' of epistemic processes. These operations on the notion of immortal life are a great example of semantic transmattering.

The immortality of cells within an organism means cancer, a deadly disease. It marks the moment when a finite organism faces its molecular infinity. Cells dividing themselves without restraint break the structure of the body and meaningfully influence its temporality by removing the future. Laboratory *in vitro* conditions, so to speak, vindicate the carcinogenic mechanisms by making them functional and wanted. Crucially, they are made observable through scaling. The understanding and appreciation of cellular immortality would not be possible without technological extensions.

Working with cells, understood as '-with' in 'I-with,' would not be possible without technological extensions, such as microscopes and measurement devices. If the scale is not manipulated, cells are only sediment in a cell culture flask and the sediment-pellet is observable mainly after centrifuging.

> It is impossible to recognize the materiality of cells when they are suspended in the culture medium. They become visible after centrifuging; the cellular sediment—pellet—remains on the bottom of the Falcon tube, while the medium forms the upper, transparent layer of supernatant (Fig. 6.2). Only seeing the cells under the microscope makes us realize that they are there, especially when using the confocal microscope. Looked at in this way, they have specific shapes, sizes, and particular movements. In the routine of their work, scientists are not deeply sensitive about every cell line they handle. However, they often say 'my cells,' meaning their work material, because generally, tissues must not be derived from their bodies. Perhaps, this regulation somehow helps them to be detached from the material they examine. My situation is different because I work with my own cells, planning to kill them. At the beginning, I had a peculiar feeling of splitting myself. I was not sure if what I held in my culture flasks was still me or not me anymore. Some of the biologists were curious about this experience of mine. (7 March 2016)

Technological extensions were fundamental in dealing with life and death in laboratory conditions. Microscope observation is not only a method of cell examination; it is also a deadly intervention for them. They are taken from the safe environment of the incubator—a technological body—and often they are not allowed to return there when the observation is done. Contact with the 'outside' makes them dangerous for the other cultured cell lines living in incubators. Having gone through microscope examination, cells become biohazardous waste. As stated before, it is much easier, emotionally, to kill a cell than to kill an animal, but observing dying cells in high resolution under the confocal microscope triggers emotional reactions, whether one is a biologist or a liminal practitioner.

> Observations of my cells under a confocal microscope gave me an opportunity to feel a particular connection with them and their interdependence at

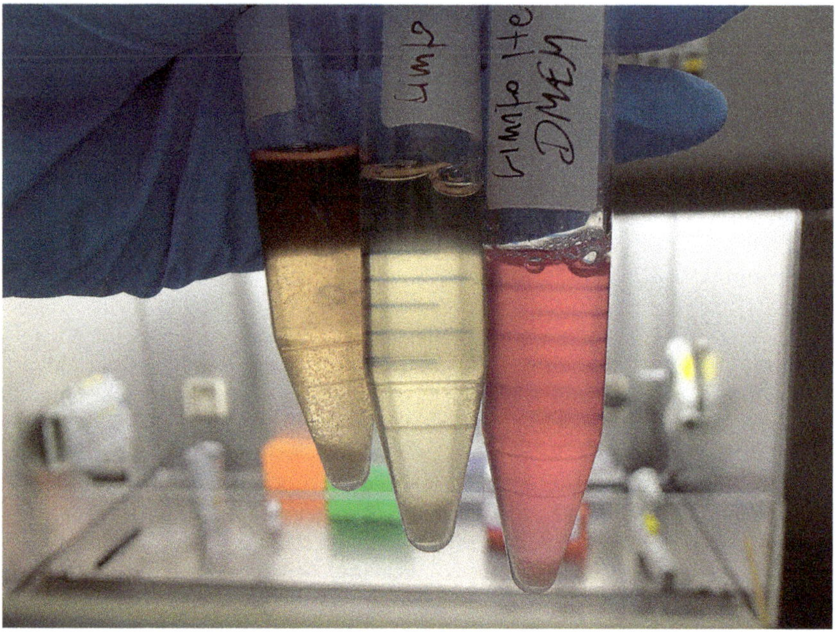

**Fig. 6.2** Cell culture after centrifuging, pellet and supernatant. Photo: Karolina Żyniewicz

> the same time. They took anthropomorphic shapes that evoked sympathy of sorts. Naturally, humans anthropomorphize everything (Fig. 6.3). Cells got a culturing shock when transported between buildings. They are pretty sensitive to changes in environmental parameters, temperature, and $CO_2$ levels. It was strange to see my cells in necrosis, vomiting their content out. I understood then how badly I tortured them. I saw them more like independent beings than elements of my body. Specimen drying is another deadly factor to cells. I had to rush to take pictures. Observing lymphocytes is tricky because they are non-adherent cells floating in the medium; their constant movement makes taking photos difficult. (16 May 2016)

Since I had learned that there were two kinds of cellular death, the main goal of my experiments was to induce apoptosis, which is amazingly programmed in the cellular genome. Apoptosis is an exquisite and clean way of dying as everything happens inside the cell, resulting in its general disappearance. Unfortunately, it was not easy to trigger this exact process.

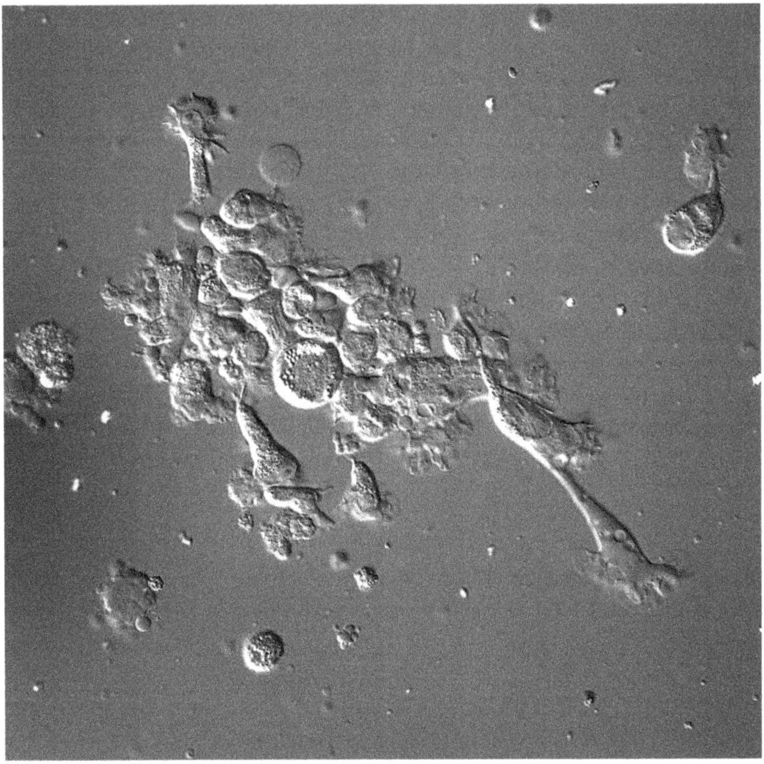

**Fig. 6.3** B lymphocytes seen under a confocal microscope. Laboratory of Confocal Microscope, Faculty of Biology, University of Warsaw. Photo from Karolina's Żyniewicz archive

On the whole, it is far easier to cause necrosis, a rapid and spectacular reaction of cells to external factors. Necrosis happens faster than apoptosis; it is messy, because the cell membrane cracks, and everything that the cell has inside spills out. An observer can almost feel pain when looking at this radical destruction of a cell. This is an empathetic response. Natasha Myers found something similar when studying a protein crystallography laboratory.

The head of the laboratory where Myers was doing her research talked about embodied empathy with molecules, which was also associated with aesthetic impressions. Describing the modelling process, she was gesturing a lot. She was like a molecule (Myers 2008). She also confessed that, when looking at a strangely shaped, imbalanced, and distorted molecule, she felt pain, almost like an artist who is not satisfied with a composition they have created: "I get stressed just looking at it…it is like I feel the pain that molecule is in, because it just can't go like that" (Myers 2008: 165). It is an exciting example of tacit knowledge; the construction of molecules is tested on the researcher's body. I had expected to respond to seeing my dying cells in a similar manner, but my approach to them changed under the influence of the laboratory. I was becoming less and less emotional. I realized that the transmattering processes, which I had expected to be deeply moving and poignant, was being moderated by technological and epistemic mediation.

> Increasingly often, I forget that I'm working with my cells. I'm absorbing the scientific way of thinking: cells are only the material for work. The biologists sometimes ask me how I feel while killing my cells. I do not know how I should feel. (30 May 2016)

Bellacasa observes that: "Making care invisible or externalized does not make it disappear" (2017: 56). What I developed in laboratory work was a denial of care. Taking care of relations with human actors of the laboratory network, as it were, necessitated denying emotionally caring about post-human liminal beings. It was actually no game of pretending; instead, it was a form of assimilation in and to the laboratory. The laboratory context can be said to have incisively transformed my sensitivity and vulnerability. Without microscopes, I would not have had empathetic reactions. However, in large-scale images of the confocal microscope, the cells are mouse- or rat-sized, and the onlooker can see them react to a deadly factor applied in an experiment from close-up. For *safe suicide*, I conducted two kinds of microscope observations: regular checks, which were an integral element of running a cell culture (to see whether the cells were developing correctly and were in a good condition), and watching the process of the cells dying, which was pivotal to the project idea and its gallery presentations. This combination implies that care has different facets.

The main aim of the first stage of the *safe suicide* project was to induce apoptosis in my immortalized B lymphocytes. It was about proving that even immortal cells were killable. Cellular immortality in the laboratory context means something else than outside of the experimental area, but the reversal of eternal existence is meaningful everywhere. It is death—the termination of biological actions.

The fact that apoptosis is called the suicidal death of cells was not the only reason why I decided to make it part of my project. I found apoptosis intriguing in and of itself. The term comes from Greek, where it denotes 'falling off,' as in the 'falling off' of leaves from a tree. The biological literature tends to describe apoptosis with recourse to anthropomorphic metaphors, such as cellular suicide, a family of caspases, and an organized community of cells. Unlike necrosis, which is a rapid reaction to a substantial injury, where the cells swell and burst, apoptosis is ordered in time and can be seen as a relational process. First of all, a cell is a part of a more extensive cell community. Every cell is an agent influencing the other cells that form organisms. As a process of self-elimination, apoptosis commences when a cell is no longer needed or poses a threat to its environment. While necrosis is a disorder that affects the dying cell's environment, apoptosis is clean and elegant and has a kind of dignity to it.

During apoptosis, the cell's cytoskeleton collapses, and DNA is fragmented. An actant family of enzymes inside the cell, called caspases, interacts to conduct the destructive process; they knock each other over one by one. The action initiated by procaspase is called a caspase cascade. A signal outside the cell can activate the initiator through a death receptor, such as Fas protein. Once it starts, apoptosis is both unstoppable and irreversible (Alberts et al. 2002). Common depictions of apoptosis are so anthropomorphized that the process seems to be a deliberate and intelligent operation. There are experimental methods of setting off apoptosis, for instance: causing a cultivating shock by changing temperature and medium quality and the administration of certain substances (e.g. doxorubicin). However, calculating the volume and strength of the trigger with proper accuracy is difficult. Some actions may be too weak, and some too strong, causing necrosis rather than apoptosis. Paradoxically enough, if causing death as such is easy, not all kinds of death are equally valid. Even if a killing experiment went well once, it does not mean that it would be easy to repeat it. Cells can react to common triggers in new and unexpected ways. Their response to the poisoning substance can be different every day. Briefly, they can be more or less killable.

> I came to the laboratory without any particular agenda. I was curious if something new had happened to my cells. It turned out that coming over was worth it. The changes at hand surprised not only me but also all my

> scientific colleagues. They said that they had never seen cells dying in that way before. The cells looked like they had stopped midway through mitosis. That was why we collectively decided to do nuclei staining to check for DNA damage. Some cells started detaching from the bottom, so it was crucial to determine whether to dye my cells alive quickly. Another possibility was fixing them and observing them the next time I was in the laboratory. That was a risk to take. I did not have enough time to watch the cells alive, so I decided to keep them for later. I was doing my best to avoid washing my cells away. In the end, fortunately, I managed to fix enough of them. (29 March 2017)

The analysis of the fixed cells did not explain why they had died. It was an example of a laboratory artefact. Theoretically, the main aim of my project was achieved, but I have never decided to incorporate those results into any presentation. That was simply beyond knowledge and laboratory procedures, and neither the laboratory team nor I could make any use of that.

Besides intentional experiments, the very proximity of a liminal practitioner, or a regular biologist, can be a deadly factor for cells. As Barad warns: "Touch is never pure and innocent. It is inseparable from the field of differential relations that constitute it" (2012: 215). Every touch, even an indirect one, causes a response. The human body, its physical border always exposed to the external environment, is not sterile; putting it differently, it is too dirty for the requirements of the laboratory and can cause contamination or infection. This is why only the gloved hands, sterilized with ethanol, can enter the laminar flow cabinet. If a scientist or a liminal practitioner is sick, has the flu, or a runny nose, complications ensue. The general question is if this poses an increased risk of infecting tissue cultures. At least, this was a problem that emerged during my laboratory practice. The point is that cells must have their medium regularly changed, and this procedure must not be suddenly interrupted due to the sickness of a scientist/liminal practitioner. It is a matter of imagination and responsibility. Everybody needs to calculate the risk. Sometimes, it is simply better to ask someone else in the laboratory to keep feeding the cells for a while.

> I have been struggling with a strange cough for the last few days. That made me concerned two days ago, when passaging my cells, concerned about their safety. The smell of ethanol irritated my respiratory tract, causing me to cough even stronger. I had to go to the vestibule for a while to avoid

> coughing close to the laminar flow cabinet. My imagination started to plot bad scenarios of an infection in my cell culture. Going to the laboratory today, I had a bad feeling that I would find all my cells dead. Even worse was my imagination about infecting the cultures run by my colleagues. (4 May 2017)

The feeling of being a dangerous stranger kept me company throughout my laboratory work. In conditions of sterility, my body, which was the 'home' of my cells, became the abject to avoid, or, to put it more precisely, my relationship with my cells was a relation between actants dangerous to each other.

Planning experiments for *safe suicide*, I was unaware that many basic measurements and protocol tasks were lethal to cells. At a certain point, I became confused about whether I should consider random procedural killings as something useful for my project. I decided that only the cells I intentionally killed would be used in microscope observations. That made me think about how variously death was understood in my project. I thought about what I should do with the cells that had died because of my mistakes in cultivating or transporting them and those used for counting or removed as looking suspicious. In the end, it did not matter how many cells I had killed in my project.

There is something perverse to taking care of cells in certain experiments. For instance, preparing immune specimens requires being very light-fingered, accurate, and calm when following the protocol. I remember being taught how to do that by a biologist.

She was tiny, and all her moves were elegant and precise. The process with its rhythm hypnotized me. This beautiful craft was a craft of killing.

> The preparation of immune specimens (Fig. 6.4) follows certain protocols. It requires excellent meticulousness, delicacy, and manual skills. The process itself is a kind of torture to cells, causing their death. From the moment of fixation in PFA on, delicacy is intertwined with cruelty. (12 May 2016)

To isolate something from a cell or to stain its elements, one must kill the cell and violently devastate its body. It is difficult to estimate whether more cells have died in planned experiments or as a result of random mistakes. Some of them have had to die for purely practical purposes, such as testing the best concentration of a substance or terminating cultivation before a holiday break or a trip. I believe that this 'practical killing' was

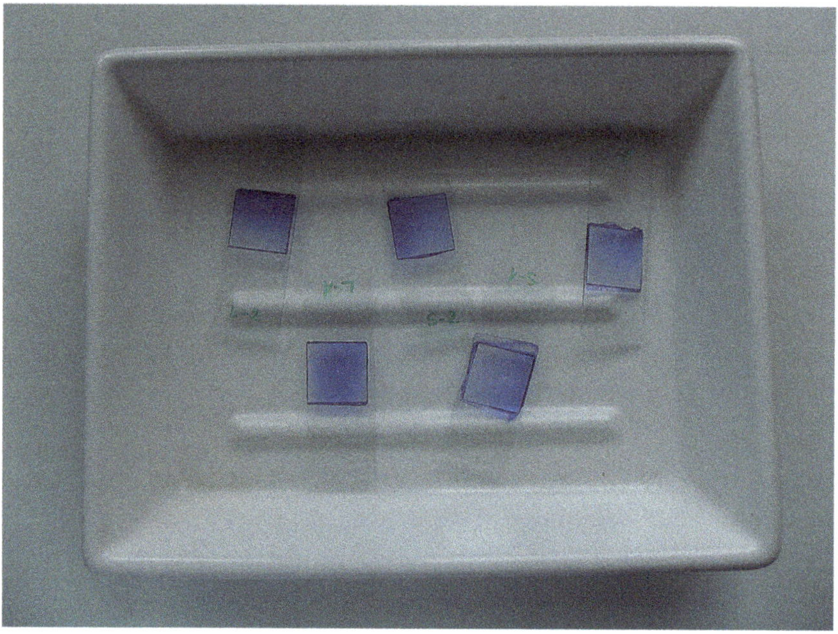

**Fig. 6.4** The preparation of immune specimens. Photo: Karolina Żyniewicz

emotionally the hardest to process, and intentional killing for *safe suicide* was usually more bearable.

The status of cell cultures is essential for the killing actions. A laboratory cell line started from the human body consists of human cells, but they are not human beings; rather, they are post-human beings and exist in the state of laboratory post-nature. They are body fragments disconnected from the body, but retaining their function. This ties in with Kristeva's definition of the abject, as an element that is relegated outside the order, rejected from the whole of the body. Once integral parts of the body, abjects are not so anymore. Kristeva's (1982) examples of the abject, such as hair found in the soup or in a sink, cut nails, and (primarily menstrual) blood, all undergo a transformation from being essential and valuable to being disgusting and rejected.

In this sense, the tissues kept alive in laboratory conditions can also be construed as the abject. This might make killing them easier. They cross the borders of the body. However, I do not think that cell lines spark the same revolted reaction as the other biological materials listed by Kristeva. One reason for this difference can be that repulsion is a primary emotion. This was how the ancestors of humans reacted to spoiled food or animal remains. As access to cells outside the body is a rather recent human invention, response to them is not shaped by primary impulses. The human brain has to adapt to quite a new situation. The fact that cell cultures remain alive in *in vitro* conditions is essential. Biologists are not disgusted by the cells with which they work, even though some of them do not like the moment of cell isolation from blood, urine, or muscles. It is thought of as a dirty stage of work. Isolated cells are not outside the order; rather, they are transferred from one order to another. The laboratory is associated with cleanness and sterility. Post-liminal beings are purified from the abject quality of the body. The mandatory cleaning that takes place after cell isolation is a symbolic obliteration of the traces of the body. The abject has a symbolic dimension to it as a symbol of crossing boundaries. Laboratory post-liminal beings cross boundaries twice. On leaving the body, they are placed in the liminal area of the laboratory. Then, they become abjectual when they lose their function in research and become biohazardous waste.

Being-with, contaminating, and being contaminated, as understood by Haraway and Tsing, are neither positive nor negative, and they are relational foundations of existence, sometimes lively and sometimes deadly. Culturing my own and HeLa cells together on the same plate was a vital experiment as part of my *safe suicide* project. Symbolically speaking, it was also underpinned by the idea of an embodied meeting with Henrietta Lacks. That idea was ridiculous from the experimental point of view, mainly because both cell lines demanded different cultivating conditions. Specifically, HeLa cells are adherent cells cultured in the Dulbecco's Modified Eagle medium (DMEM), while B lymphocytes are non-adherent and grown in the Roswell Park Memorial Institute (RPMI) medium. What was crucial for the experiment was that both lines were immortal. Knowing that HeLa cells were highly contaminative for other cell lines and that they grew fast in all conditions, I wanted to hold a 'competition,' whereby I expected HeLa cells to curb the growth of my B lymphocytes. The meeting with HeLa cells could be a way of killing my immortal cells. It was complicated to determine the best medium for cultivating the two cell types together. Give this, I prepared three plates: one with DMEM, one with RPMI, and one with a mixture of both mediums.

Surprisingly, the joint cultivation went well on each plate, and both cell lines developed at more or less the same rate. I did not investigate if these cell lines changed each other genetically, but I suppose that they mutually influenced each other's growth. The conjecture that HeLa cells would kill—or at least reduce—the vital functions of my immortalized cells was disproved. The mixed culture was incorporated into the project, and I initiated cellular death in various experiments, observing how HeLa and my cells die-with.

As already mentioned, not all research solutions come up on time. Sometimes, a piece of knowledge appears far too late to be applied. Testing for Mycoplasma was an example of such a bad timing in my research. The test should have been done at the beginning of the cultivation. Doing the test is so obvious for biologists that my scientific advisor forgot to mention it.

Anybody can be an inadvertent Mycoplasma carrier, so testing for it before donating tissues is essential. Cell cultures can also be infected in the laboratory. Research performed on a cell line infected with Mycoplasma is not regarded as reliable. The infected line can also very easily infect all other cultures grown in the laboratory at the time. The bacteria are antibiotic-resistant, so getting rid of them is a major problem, and it is usually easier to eradicate the infected cell line than to cure it. In *safe suicide*, the Mycoplasma test was done at the end of cultivation before putting the storage tubes in a tank with liquid nitrogen. That put at risk all the previous work and the future of the cultivation. The bacteria were a common enemy to me and my cells. I could lose all my results, and my cells could lose their life. I would have had to destroy the whole cell line, if it was found to be infected. That would have been an unexpected and unplanned finale of my project.

> The result of the first test was not promising. It looked like my cells had stowaways on board. We decided to repeat the test, this time on a portion of cells frozen earlier. If the second test came out positive, I would have to test all my cultures. People in the laboratory were surprised that this was such a big deal for me, but I was worried about losing such a substantial amount of cells. They were joking that it would be a perfect final for my project, a massive burial of my cells in hypochlorite. (28 June 2017)

The biologists persuaded me to incorporate this unusual situation into my project. For a liminal project, something like that could be, in fact, an intriguing punch line. Regular scientific research often involves losing time and money and starting everything from scratch. My work ethic did not allow me to accept random death, and I insisted on determining when and how to kill my cells. Fortunately for my plans, one result is not enough to annihilate a culture. A covert action of non-living actants always has to be taken into account.

> This time I decided to do my best to be properly focused on doing the Mycoplasma test. First, I showed people in the laboratory how I had done the test before because they suspected I had made mistakes. They were astonished that the combination of starters noted in my lab notebook was F1F2 and R1R2, instead of F1F2 and F2R2. That looked like a fundamental mistake that could lead to wrong results. There was a visible band in the control section. That meant that someone had probably contaminated the starters before. This time I consulted every step and did the primary dilution completely focused. I had trouble preparing mixtures. When I started preparing the second, I realized that I was unsure if I had added starters to the first. K. had entered the room for a moment, asked me a question, and briefly distracted me. Instead of adding mixtures to test tubes, I had only added water. I had to repeat the procedure. M. supported me in setting PCR to control if I was doing it correctly. I was not sure if I had done everything correctly, but M. offered to control all my actions the following time.
>
> (…)
>
> Because the band was also visible in the negative control section, M. decided to try it in her gel. She suspected that Mycoplasma could have contaminated the negative control.
>
> (…)
>
> I asked M. in the afternoon about the result of her trial and found out that she had also got a positive result. This means that I may have contaminated something when preparing PCR or that the DNA donor was a Mycoplasma donor. (date not noted)

The investigation into the Mycoplasma predicament was very complex and stressful. The last test, done using fresh starters, showed that there was no Mycoplasma in my cell line. That was a huge relief. The episode also vividly illustrated how many questions must be asked when something went wrong. Looking at the incident from a different perspective, I could accept the agency of Mycoplasma and acknowledge it as transmattering by a random encounter.

As explained in the chapter on Science and Technology Studies (STS) research, the value of laboratory liminal lives is dictated by how useful they are in given experiments. Knorr Cetina argues that, in molecular biology, life is viewed as a self-reproducing biological machine (1999: 157). This way of thinking can promote biologists' emotional detachment from post-liminal beings and model organisms. I also did not expect to develop such a distance. Caring about my cell lines had a realistic side to it. They were hired to play a role in my theatre play.

In her depiction of liminal biological beings, Squier stresses that, as a society, we neither come in touch nor engage with them often or closely. They are mostly known through mediations, translations, and representations. Very few people have access to laboratories to see what tissue cultivation looks like. That can spark imaginary notions that scientists directly interact with the 'liminal aliens.' Meanwhile, sterile work hardly involves any direct meddling with the biomatter under examination. Wet work, in general, is embodied research based on multiple tasks, but any direct contact of tissues with the borders of the human body, the naked skin, is strictly prohibited. The skin is exposed to the external world, which is dirty and polluting. When isolating fibroblasts from my skin biopsy, I had to very carefully remove the epidermis, which usually swarms with various bacteria. The surface of the human palm, touching elements of the unclean environment, is similarly infested with bacteria. Consequently, to work in a laminar flow chamber, one must wear disposable gloves. Glass protects the sterile interior space from everything that is exhaled from the human body while breathing. Breathing is even more contaminating than physical touch. All activities involving cell culture are performed by means of disposable pipettes. This can be considered touching without touch, or, as Bellacasa suggested: "caring without touch. It is an example of when caring means not touching but instead keeping a certain distance" (2017: 5). However, in Barad's perspective, every measurement is a matter of touch (2012: 208). Any careless move can be dangerous to cells; for instance, if one randomly touches the pipette top (which absorbs the medium with cells), even when wearing gloves, one can contaminate it, and the same concerns touching the neck of a culture flask. The fundamental thing to learn in the laboratory is that bare hands cannot as much as be put inside the laminar flow cabinet. There is always an artificial barrier between one's palm and a test tube, a culture flask, a petri dish. It was when thawing my cells that I experienced the closest contact with them:

> The defrosting of cells is in a sense magical. The optimum temperature for cell life, which is set as the constant temperature in the incubator, is the average temperature of the human body: 36.7° C. When A. suggested that I warm the freezing tube, holding it in my clenched palm, I felt a tenderness

that my body was giving to the elements that had been derived from it. (3 March 2016)

The closeness described in this episode involved two artificial protecting layers: the plastic walls of the test tube and the rubber gloves I was wearing.

Discussing the connection between care and touch, Bellacasa concludes: "Care is not about fusion; it can be about right distance" (2017: 5). I believe that our care is indeed about fusion as long as our cells are integrated with our bodies. However, the moment that cells are isolated from our organism, it becomes necessary to keep a certain distance. Getting too close is reckless and irresponsible. Meeting with post-human liminal beings is meeting with strangers, even though they were once integral parts of the human liminal being's body.

While still integrated with the body, cells are not easily visible, but we do not try to observe them for themselves; rather, we examine the general condition of the body. Isolation can be viewed as a moment when cells are neutralized. As they are outside the body, they cannot harm it by any uncontrollable development—for instance, morphing into cancerous cells. Cells growing on a plate are mainly dangerous to themselves.

My caring about the cells taken from my body was of a palliative nature. I was preparing them for death, and they had to be in a good shape for the sake of microscope observations. The metaphor of palliative care was even more pertinent in the second stage of *safe suicide* because then I observed the process of cellular ageing. Rather than intentionally causing death, I was waiting for it to occur naturally. The *in vitro* environment made it possible for me to experience all the unavoidable body processes, which tend to scare humans, at an accelerated pace. Taking care of my ageing and dying cells, as it were, prefigured the care I would like to experience in *in vivo* conditions.

Doing research based on cell lines is a matter of responsibility for the quality of their cultivation. Besides that, culturing also means response-ability to all human and non-human or post-human actors of the laboratory network. I felt responsible for planning my transmattering, and I had to respond to its disobedience. As Haraway observes: "Shaping response-abilities, things, and living beings can be inside and outside human and non-human bodies, at different scales of time and space" (2016: 16). The response-ability of a liminal practitioner is quite individual and independent; meanwhile, their responsibility is shared with a scientist/scientist, their collaborator/s in liminal project production. The scientific partner is responsible for the liminal practitioner's responsibility, and while the former's control influences the latter's response-ability, this control is far from total, especially regarding emotional reactions.

> I took a look at all the three Petri dishes under the microscope. At the same time, in a multitasking style, I was preparing the laminar flow cabinet and the medium. A. also took a look at my plates, and she said that the fibroblasts were ready for passaging on one of them only. In the remaining two Petri dishes, I had just changed the medium (Fig. 6.5). In one of them, the biopsy had peeled off, so I had to place it on a fresh plate. A. was supervising everything I was doing. Firstly, I had to remove the old medium using a Pasteur pipette. Then, I had to slowly drop trypsin onto the spot with a biopsy sample. Trypsin allows cells to peel off. The best way is to peel off the migrating fibroblasts without peeling off the biopsy piece, but this is very tricky. Trypsin needs at least 10 minutes to work, and I used that time to work with the other plates. I wanted to see the process of cells detaching

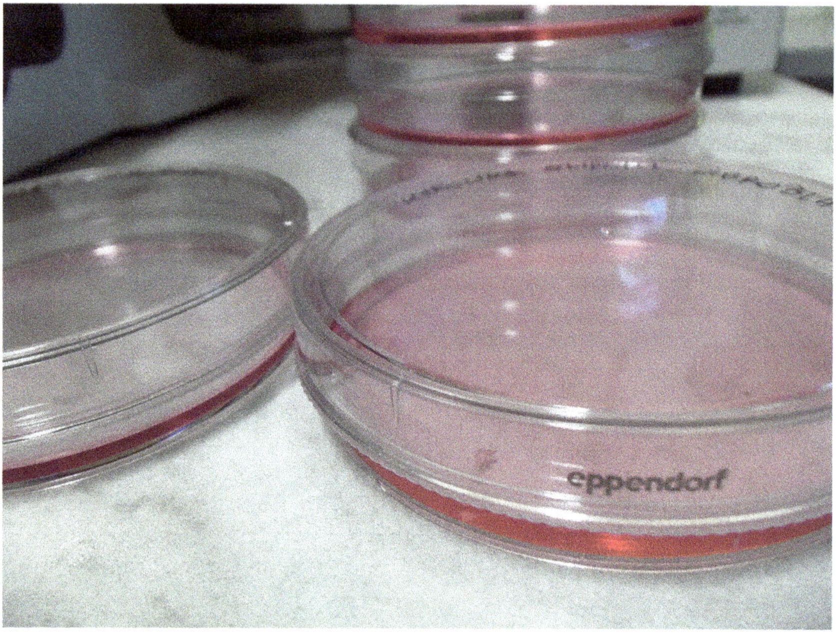


**Fig. 6.5** Fibroblast culture. Photo: Karolina Żyniewicz

from the bottom, so I moved the plate under the microscope. A. suddenly screamed: Don't! I got scared and felt like a kid. She explained to me that the rapid movement of the plate also caused the moving of cells. That could make finding them under the microscope very difficult. She offered to do it herself that time around, teaching me how to do it smoothly enough. She slowly dropped a small amount of the medium by contracting and releasing it several times; then, she transferred the whole liquid to a falcon. The next step was centrifuging. After that, I moved all suspensions to a new petri dish. (4 November 2016)

Killing and cruelty, which I have presented as essential elements of laboratory research, including mine, are also responses. Usually, only a particular component of the cell is needed for experiments, and this element must be isolated from the cell, just like the cell was isolated from the body before. The difference is that the body is not destroyed when some of its parts are separated, whereas cells are usually destroyed by having their mitochondria or nuclei isolated from them. It is necessary to embrace all forms of response-ability and different stages of transmattering. It is not only about accepting our responses but also about accepting how others respond. The limits of acceptance depend on one's personal and professional ethics:

M. switched the laminar flow chamber off, saying that his culture (which is vast) is to undergo lysis. I asked what it meant. He explained that lysis was a process of destroying cells in order to isolate mitochondria. He uses a particular buffer that swells the cells by making them absorb water. In the end, they crack from the rising pressure inside. Only mitochondria are able to survive the process. It was a bit scary, so I thought of the procedure as a cellular Holocaust. (10 November 2016)

Barad insists that: "matter is a condensation of response-ability. Touching is a matter of response. Each of us is constituted in response-ability. Each of 'us' is constituted as responsible for the other, as the other" (2012: 215). In my understanding, this means that our materiality is a guarantee of our capacity to respond. Our bodies know how to respond, even though the response/action is not always optimal in terms of responsibility.

Sometimes things take the opposite course; responsibility dictates a response that appears unnatural.

> I found that the cells in one of my Petri dishes were not attached to the bottom properly. A., as my scientific advisor, suggested that I remove the plate. I had to do something against myself, pouring the medium with living cells into the sink. When it was done, I realized that I could have left the cells in the petri dish, allowing them to die at their rate. (16 December 2016)

The fact that materiality is 'a condensation of response-ability' means that there are many responsive others. Human domination is just an illusion and, if anything, working with cells only provides strong evidence for that assertion. This is often an excruciating lesson to learn. Humans are the most decisive actors of laboratory networks, but they are certainly not the most responsive ones. Response-ability is distributed equally among many various actants.

> The magic continued today. This time my cells attached exceptionally firmly to the bottom of their Petri dish. There were so many of them that it was hard to see any free space. My consultation with A.: Me: 'Why did my fibroblasts suddenly get round-shaped and do not look like fibroblasts anymore?' A: 'They can be so dense that it is hard for them to attach to the plate bottom properly. You need to passage them.' I followed the suggestion and prepared my cells for passaging. First, I poured 3 ml of trypsin to help my cells to detach from the petri dish bottom. I knew that the detachment process took some time, so I got a bit stressed that E. would have finished her work, closed the sterile workroom, and given the key back at the front desk downstairs by then. During the next consultation, A. suggested that I knock gently on the petri dish bottom. I also started to rinse the bottom with a mixture of trypsin and PBS, pipetting it with more pressure. All in vain. I began to panic, trying to figure out what I should do; I could not even pour the content of the plate into the sink before the cells were detached. In the end, I decided to remove the detaching liquid. I supported the cells with some fresh medium, leaving them in the incubator. It is hard to predict what I will see on Monday. Probably, the cells will be so dense that they will begin to die. (16 December 2016)

Responsible for the quality of research and biosafety in the laboratory, scientific advisors tend to decide that suspicious cell cultures must be destroyed. From the economic point of view, this is much easier and more reasonable than continuing to look for reasons why cells have responded in a particular way. Sometimes, investigating abnormal cell responses is nevertheless undertaken, and it is a matter of weighing pros and cons.

> I was afraid to check on my cells and see their condition. Fortunately, A. had finished her holiday, and she could take a look at my Petri dishes. I made a joke, saying that she had never seen such dense cells. She responded by joking that she had already seen almost everything. Still, my cells were indeed able to surprise her. She could not say what she saw under the microscope. In her opinion, my cells did not even look like fibroblasts. This uncertainty forced us to pour the contents of the plate into the sink. It is always a challenging task for me. I could see the cell sediment on the bottom even without using the microscope. It was evident that some of them were already dead. I started to wonder if it would be possible to save them, but I had to respect the decision made by a much more experienced person than me. (9 January 2017)

Wrong responses in the scientific field generate real cost, in terms of not only the precious research time but also money. That is why being responsible for research often entails limiting certain forms of response.

When developing my *safe suicide* project, I worked with cells unrelated to my body as well. I have already mentioned cultivating my cells together with HeLa cells when the latter became incorporated into the project design. Besides working on my liminal vision, I also worked with anonymous cells during my internship. It was also a kind of experiment to see whether my approach to cells not related to my body would be very different from my attitude to my own cells.

My main task as an intern was to grow a mysterious cell line which was self-immortalized (so cancerous). They were muscle cells, but I did not know who the donor was. When I began to take care of them, they became mine, in the sense of being my research material for which I was responsible. I adopted them the way biologists did.

Relations with cells derived from someone else's body or bought from a tissue producer are even more stressful than relations with cells genetically related to one's body. Whatever one does individually affects the research conducted by a laboratory team as a whole. I knew that the foreign cells required the same care as my cells. In *safe suicide*, my cells were mine in a double sense as they came from my body and were also my research material. What all my relations with cells had in common was fascination. I appreciated watching the cells that I did not know:

> Today I saw the biologists watch time-lapse videos. I could not focus on anything else, seeing cells stained red and green. There was random music on the radio, and the cells looked like they were dancing. The biologists

> observed interactions of the cells. I was curious about the research objectives, but they told me they needed to find it. Sometimes, the cells touched each other; sometimes, we had an impression that they were fighting. (…) W. showed me a video recording the movement of cells attached to the bottom of a Petri dish. Even when attached, they were never static. (1 August 2017)

In *safe suicide*, my foremost job was to kill my cells in particular ways, as an intern, I had quite an opposite task to do since, rather than killing the cell line unrelated to my body which I had been given, I was supposed to keep them alive and in a good condition. Nonetheless, I had to accept there some cells had to be eradicated just because there were too many of them to work with. The moment of death was not something to pay attention to. All this was down to cold deliberation:

> My internship involves some moral dilemmas. To observe the development of cells, I need to keep the same amount of them as the point of reference. I asked my laboratory colleagues what I should do with the surplus of cells, and they told me that they had no plan of using them. This means that I have to kill the redundant cells by pouring hypochlorite over them. Like many times before, I felt terrible killing healthy cells. (3 August 2017)

This begs an important question: Was my work with the cells unrelated to my body and my projects a form of transmattering as well? My answer is that the main thing that was transmattered in this case was the work purpose. I performed a range of operations during my internship, but none of them crossed the limits of regular biological research. I put myself to a test to see if I could work in a laboratory having no transmattering intention.

## References


Agamben, Giorgio. 2009. *The Signature of All Things: On Method*. New York: Princeton University Press.

Alberts, Bruce, Alexander Johnson, Julian Lewis, Martin Raff, and Keith Roberts. 2002. *Molecular Biology of the Cell*. 4th ed. New York: Garland Science.

Barad, Karen. 2012. On Touching—The Inhuman That Therefore I Am. *differences: A Journal of Feminist Cultural Studies* 23 (3): 206–223. https://doi.org/10.1215/10407391-1892943. Accessed 31 July 2022.

Deleuze, Gilles, and Félix Guattari. 2017. *Anty-Edyp: Capitalism and Schizophrenia*, vol. 1. Translated by Tomasz Kaszubski. Warszawa: Wydawnictwo Krytyki Politycznej.

Haraway, Donna. 2016. *Staying with the Trouble Making Kin in the Chthulucene*. Durham & London: Duke University Press.

Hayflick, Leonard. 1965. The Limited In Vitro Lifetime of Human Diploid Cell Strains. *Experimental Cell Research* 37 (3): 614–636.

Jiang, Lijing. 2010. "Alexis Carrel's Tissue Culture Techniques." *Embryo Project Encyclopedia*, June 18. https://hdl.handle.net/10776/2015.

Knorr Cetina, Karin. 1999. *Epistemic Cultures. How the Sciences Make Knowledge*. Cambridge & Massachusetts & London: Harvard University Press.

Kristeva, Julia. 1982. *Powers of Horror: An Essay on Abjection*. New York: Columbia University Press.

Landecker, Hannah. 2000. "Immortality, In Vitro: A History of the HeLa Cell Line." In *Biotechnology and Culture: Bodies, Anxieties, Ethics*, edited by Paul Brodwin, 53–72. Bloomington, IN: Indiana University Press.

Landecker, Hannah. 2007. *Culturing Life. How Cells Became Technologies*. Cambridge & Massachusetts & London & England: Harvard University Press.

Malabou, Catrine. 2004. *The Future of Hegel: Plasticity, Temporality and Dialectic*. London: Routledge.

Malabou, Catherine, and Kate Lawless. 2016. "The Future of Plasticity: An Interview with Catherine Malabou." *Chiasma: A Site For Thought 3 (1)*: 99–108.

Myers, Natasha. 2008. Molecular Embodiment and the Body-Work of Modeling in Protein Crystallography. *Social Studies of Science* 38 (2): 163–199.

Puig de la Bellacasa, María. 2017. *Matters of Care: Speculative Ethics in More Than Human Worlds*. Minneapolis and London: University of Minnesota Press.

Skloot, Rebecca. 2010. *The Immortal Life of Henrietta Lacks*. New York: Crown.

Squier, Susan Merrill. 2004. *Liminal Lives: Imagining the Human at the Frontiers of Biomedicine*. Durham and London: Duke University Press.

CHAPTER 7

# *The Last Supper*: Genetic Transmattering and Self-Digesting

*Today, we are learning the language in which God created life... With this profound new knowledge, humankind is on the verge of gaining immense, new power to heal.*
*Bill Clinton, 26 June 2000*

Bennett's concept of vitalism is crucial for understanding my experiences during the development of *The Last Supper*. Her vision of material assemblages consisting of several independent elements inspired me to carry further my physical transmattering commenced in the *safe suicide* project. My cells were living beings, performing their metabolic processes. I decided to penetrate into their bodies and isolate their lively essence—the genetic material. I wanted to use its plasticity and lively potential to foster a special bond between my body and the body of yeast. The protein being which is at the same time a code, a form of language, had to be transported between my body and the body of yeast by living vehicles: bacteria.

Producing transgenic yeast was at the core of the project, but it was not its primary or ultimate goal. The transformed yeast organism was supposed to cause fermentation and produce alcohol, a foodstuff. The project relied on Bennett's idea of edible matter as an 'assemblage converter.' People drinking (intentionally unclarified) beer brewed in the project were envisaged to absorb yeast remains, while my own gene had earlier been absorbed to be the yeast. In this way, different elements of the assemblage


K. Żyniewicz, *The Body as Matter in Art and Science*,
https://doi.org/10.1007/978-3-032-13665-7_7

would be blended. Crucially, the trajectory of *The Last Supper* started from the human body and ended in another human body. A gene extracted from my body did not add anything new to the consumers' bodies because they had the same gene as well. The project pivoted on the transfer of the gene from one person through and by temporary living vessels to another person. This trajectory was made possible by the scientific discovery of how genomes could be read and translated into a language understandable to humans.

The words uttered by Bill Clinton in the wake of the success the Human Genome Project had in revealing the content of the human genome and comparing it with the genomes of several other species. The genome mappers found out that 50% of the human genome was a repetitive junk sequence; 10% was made up of one million scattered copies of a single sequence which they called the Alu element. The role of junk DNA has not been fully explained, yet continues to pose a challenge to biology. Reading the genomes, though a breakthrough accomplishment in and of itself, was only the basis for experiments with transforming them, with this research stirring a lot of doubts and fears. Junk DNA is referenced by Bardini in his study of junk art. Biological junk has its reflections in culture, including junk art. *The Last Supper* was not dedicated to examining junk DNA, but it was an example of junk art. I believe it can be dubbed genetic storytelling.

Having found out about the genome sizes of various species, scientists became curious as to what was the smallest genome necessary to keep the organism alive. In 1995, the Institute of Genomic Research (TIGR) sequenced the smallest-known bacterial genome of *Mycoplasma genitalium*, consisting of 580,000 base pairs. That was a starting point for experiments carried out by Craig Venter to see how big coding fragments of *M. genitalium* could be removed without stopping its life. Venter removed 100 protein-coding genes, ending up with the 'minimal genome' of 382 genes. The metaphor of a kid destroying a toy to see how it works aptly captures Venter's undertaking. He disassembled a living mechanism and then put the components back together, excluding certain 'screws.' The experimentation on actually existing bacteria inspired Venter to construct a synthetic genome (Venter 2008), which resulted in synthesizing *Mycoplasma mycoides JCVI-syn1.0*. Proud of his achievement, Venter announced that: "This is the first self-replicating species on the planet whose parent is a computer" (Watson et al. 2017: 276), a statement proclaiming that humans could finally create life based on computer data. In fact, Venter did not create a new species from scratch; rather, he replaced the existing *Mycoplasma* genome with the one he had synthesized, which enabled him to control one aspect of life. Experiments of this kind must be accurately explained. Otherwise, the lack of knowledge makes the possibility of genetic transformations scary, especially to religious people, who believe that the creation of life is the sole domain of god/s and that humans should not even try to seize the control of life. I relied on Venter's way of thinking and attitude in my project, testing social fears related to Genetically Modified Organisms (GMOs). To do so, I needed to materialize what was first designed on the computer.

The reading of many genomes, including human ones, has shown that they share many common elements. Scientists have invented a rescue method that evinces this similarity. The method makes it possible to knock out a gene encoding a protein from the genome of one species and then replace it with a corresponding gene from that of another species. The fact that the human genome shares several components with those of other species supports the idea of the common evolutionary origin and undermines the notion of human domination and exceptionality. Genomes have been read and can be edited like a text. This development has made it possible to write different genetic stories and encouraged me to write my own one.

In *The Last Supper*, I was primarily interested in GMOs, their creation, and the general public discourse on genetic modification, especially in the food industry. I wanted to see how GMOs work as an example of edible matter and an assemblage converter. It is not my object here to detail all the pros and cons of GMOs as raised in the ongoing public debate or the broad and complex discourse it fuels. I will only address the cons of GMOs that were relevant to my liminal project. James D. Watson observes that:

> As in the case of genetically modified foods, DNA has in the popular imagination a voodoo quality: there's something scary, mysterious about it. And a lack of understanding of genetic complexities leaves one susceptible to the worst anxieties and conspiracy theories. Once people understand the issues, I hope this hesitation in making the most of a new and powerful beneficial technology will vanish. (Watson et al. 2017: 384)

This assessment ties in with my insights about the public imagination of what is happening in the laboratories. The lack of knowledge and understanding makes people perceive biotechnology almost as 'black magic.' In an interview for the *Washington Post*, Jayson Lusk, an agricultural economist at the Oklahoma State University who has studied people's aversion to GMOs since they were introduced in the late 1990s, similarly stresses the importance of the marketing uses of public ignorance. He points out that producers and sellers of bio/organic products started extensive public discourse on GMOs. They label their products as 'GMO-free,' suggesting that GMOs are the opposite of their healthy merchandise. Lusk reports an exciting experiment revealing how people react to labels. The experiment used apples with a label that said: "This is an ethylene-ripened apple," which is a common and safe process. Ethylene is a natural plant hormone that regulates the ripening of fruit. However, people objected to those apples, as to GMOs, simply because they did not know what ethylene was (Ferdman 2015). In Lusk's opinion, that was proof that people avoided new things that seemed to be uncontrollable. Michael Robin, a science writer and communication strategist, also asserts in his blog that people are busy with life and lazy in information selection (Robin 2020). Relatively few people check whether the information is credible; the rest take it unexamined and for granted. Robin also argues that we still live in tribes and dwell in bubbles, sharing the views of people who are closest to us. It takes considerable effort and mental strength to express an opinion that contradicts the one endorsed by the tribe.

In Robin's opinion, money and marketing play a substantial role in keeping the public nescient, as the investment in disinformation advertisement considerably exceeds that in sound education and science popularization. The research conducted in scientific laboratories is barely communicated to the public and its findings are not popularly known. As a result, even though, as Lusk reminds us, the scientific community agree that genetically modified crops are no more harmful than regular crops, people mainly think of GMO production as a corporate business, making huge money in often unethical ways.

Besides the knowledge and business factors, there are, so to speak, more emotional reasons for fears about GMOs. The essential question is why there is such a discrepancy between what science tells us about GMOs and what people think in the first place (Blancke 2015). One answer is that, unlike scientific knowledge, anti-GMO information is intuitively appealing, and, as such, it quickly grabs public attention. So-called psychological essentialism, which holds that DNA is an organism's essence, determining its uniqueness, also contributes to this effect. Given the essentialist premise, transferring genes from one organism to other entails changing the nature of the gene recipients; for instance, a tomato with a fish's gene will taste like fish. This mindset is fed by anti-GMO organizations, such as NGOs that flood the media with pictures of GMO monsters, as patchworks of at least two species. The 'Cisgenic' transfer of genes, that is, transfer within the same species, is less controversial. Another emotional facet of anti-GMO approaches is rooted in religion and the belief in the holy power of nature. Such convictions propel thinking that producing transgenic organisms is 'unnatural' and dangerous, resentment at 'playing god,' and worries that this will lead humanity to disaster. The image of nature as pure and innocent is coupled with disgust as a specific emotional reaction to GM food, which is considered to be contaminated by something 'unnatural' and possibly harmful. It is kind of the abject, something outside the 'natural' order.

Interestingly, the repulsed reactions mostly concern food and are as a rule not related to GMOs used in the production of medicines (e.g. insulin). The human sensitivity about eating laboratory-produced foodstuffs was intriguing enough for me to dedicate a liminal trajectory to its examination. I inquired whether GMO products were seen as abjectual, or whether they were a separate category. I created an artistic label, similar to those cited by Lusk: "this beer is humanized by a piece of my body," curious to know if it would work exactly like commercial labels. Would the drinking of transgenic beer be perceived as an act of cannibalism? Was tasting such a beverage worse than eating a soup with someone else's hair in it? The audience's/consumers' aversion could turn against the transgenic yeast producing beer or even more against the human gene intentionally incorporated into yeast organisms. The drink was contaminated with one of my genes. Rationally speaking, eating and drinking entail absorbing multiple various genes daily. I would have left more of my genes in the beer by simply sticking my finger into the glass. In *The Last Supper*, transmattering was built on narrative, on storytelling plotted around laboratory experiments.

I decided to conduct transmattering, a doable process, instead of the biblical transubstantiation. My scientific collaborator Dr Jakub Piatkowski and I wanted to transmatter my body into an element of the body of yeast and then into the ingredients of beer and bread. It took more than one year to execute the project. Besides the main scientific collaborator, the laboratory team leader (and at the same time my PhD advisor) Professor Paweł Golik was also engaged in the project. The laboratory was the place of the genetic transformation of yeast and the site of food production.

This fact was meaningful to some consumers, who perceived their drink as contaminated by the laboratory context. In scientific institutions, there are separate rooms where biologists can eat. Eating is not allowed in the experimental area. Likewise, laboratory refrigerators and freezers cannot be used for food storage. The production of beer humanized by the yeast with my gene took place in a sterile workplace environment. That could be seen as a guarantee of cleanness, but the intercontextual contamination proved a more potent factor.

Both of the projects, *safe suicide* and *The Last Supper*, are organically connected, forming two stages of the laboratory transmattering process. I compared the laboratory trajectory to playing with a Matryoshka doll. The trajectory involved going from outer to inner-more to inner-most levels and extracting from each of them a form of matter essential to the entire constellation, and at the same time capable of functioning when extracted from it. I started working on *The Last Supper* while still developing *safe suicide*. From the strategic point of view, it was easy because I just needed to navigate between two laboratories on the same floor in the same institution. At the beginning, genetic operations were my secondary occupation, compared with the time I was devoting to my cell culture. A portion of the cells that were primary post-human collaborators in the *safe suicide* project were sacrificed to obtain RNA. I had to annihilate them by administering TRIzol. Isolation was another way of killing my cells, so their end was the beginning of something new (as typical of the laboratory cycle of matter). The procedure also marked the beginning of my work with something invisible, even when inserted into the bodies of bacteria or yeast.

My genetic transmattering is a story of believing without seeing, of trusting the devices, the mathematical data, and the opinions of my scientific colleagues. This was the essential difference between the two projects. When working with cell cultures, I was able to see the effect of my action under the microscope. Genetic operations were based on designing things on the computer and then experimentally verifying the results designed. It was difficult to imagine that adding a segment to the plasmid map would subsequently affect processes taking place in bacteria. *The Last Supper* required far more specialized knowledge than *safe suicide*.

> Working with cells is within my brain's competence; genetic calculations are not. The experiments I have conducted with J. are not entirely comprehensible to me. The main problem is that I cannot imagine what is happening

during chemical reactions. Some terms and mental leaps are hard to remember. Genetic work also precisely means pipetting minimal quantities, such as one microliter (Fig. 7.1). I am afraid to do that alone because the slightest mistake influences the PCR result. (25 May 2016)

The moment that RNA was isolated from my immortalized B lymphocytes, my matter began another stage of the transmattering journey, which was based on different methods. I had to find a way of imagining the genetic operations on my DNA. Since I am a knitter, I visualized working on DNA strands as manipulating yarn threads. They can be tied or untied, and there can be knots that have to be cut off.

The aspect of caring was present throughout my transmattering journey. In *The Last Supper*, I had to learn how to care about something that was out of my perception. The transmattering operation descended to the level of genetic information that determines all the vital processes of a cell.

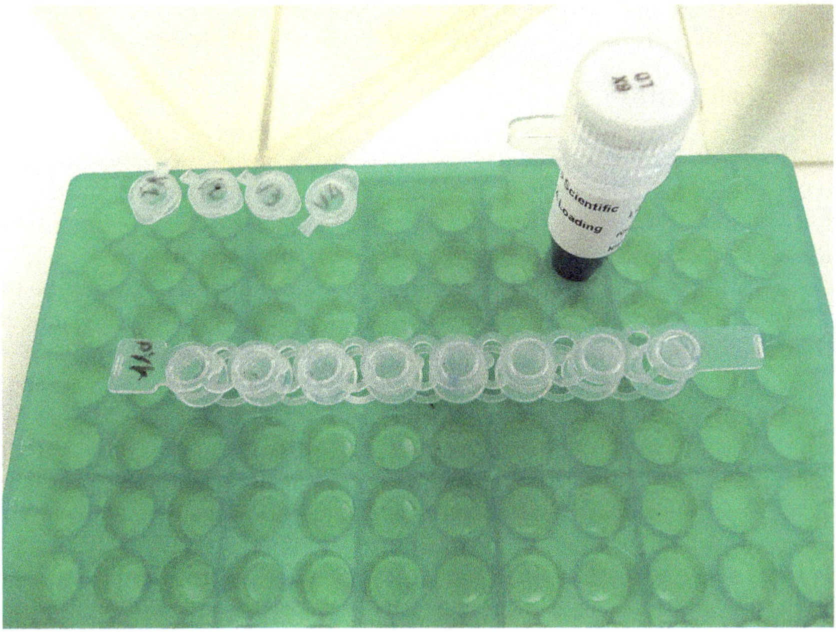


**Fig. 7.1** Preparing PCR samples. Photo: Karolina Żyniewicz

Cells as bodies and containers stopped being important. Their content became liminal in that it could be transferred to something else.

I was surprised to learn that we needed to isolate RNA, rather than DNA, from my cells, but my scientific collaborator explained the role of introns in human DNA to me. We did not want to clone introns, and mature RNA does not have introns. The plan was to do reverse transcription to obtain cDNA. I needed to imagine the situation of reverse transcription, so J. made a simple drawing for me. Later, he also used drawings to explain many other processes to me. This was useful because my scientific collaborator made genetic processes, so to speak, visible by representing them graphically.

> It is tough for me to imagine interaction between primers and the template double helix. I have even more serious problems trying to understand how overhangs work. That is why I asked J. to draw this for me. Based on a simple sketch of a single strand, two straight arrows representing primers, and two curved arrows representing overhangs, J. explained to me what we were going to do. I got lost at the point when, after adding the overhang, the starter 'stop' moved to its end. I was also confused why there was only one strand in J.'s picture. It turned out that J. had not drawn the second strand because its being there had been self-evident to him. (17 November 2016)

However, I still faced the difficulty of translating this theoretical knowledge into actual practice with physical matter. If not looked at under the microscope, DNA and RNA were visually identical as bits of sediment on the bottom of their respective test tubes. Developing the project was based on repeating certain operations, pipetting certain liquids from one tube to another, following the instructions of my scientific collaborator. I was able to see substances, transformed bacteria, and yeast colonies, but I was not expected to examine their structure and condition under the microscope. I thought it strange that the transformation tasks were not supposed to be confirmed by microscope observations. However, the point was that all these genetic operations required pipetting tiny volumes of mixtures, such as one microlitre. With such quantities, I could not see any pellet or even a drop of liquid. I had to do pipetting very carefully, believing that I had indeed sucked something in. Caring meant then gentleness and precision. In particular, polymerase chain reaction (PCR) tubes were tiny, and every ingredient of the so-called PCR mixture had to be added very carefully.

PCR was one of the primary procedures I repeated a hundred times during the realization of *The Last Supper*. I never learned how to do it on my own. PCR is a method for amplifying a particular DNA fragment. It became a workhorse of the Human Genome Project. I knew that I had to prepare a mixture, including a DNA dilution, primers, water, deoxyribonucleotide mix, a buffer solution, and DNA polymerase enzyme, but the ratios of these ingredients were always different. There was no fixed pattern that I could follow. Every time, my scientific collaborator established the proportions, saying: "So today you will mix…" and wrote the recipe down in my lab notebook. My role was only to pipette all the ingredients gently and without mistakes. I felt like a laboratory technician. I could see the ingredients (in tiny test tubes), but I could not grasp their meaning in the experiment.

> I was happy like a kid today because I did my first PCR. I went to the cellular laboratory to announce my big success. They asked me, laughing at my enthusiasm, what that PCR was for. They laughed even more when I confessed that I had no idea. (30 May 2016)

Even setting the thermocycler was too complicated for me. Even if the machine's programmes were relatively simple, I had problems deciding which programme should be used. I felt so responsible for the DNA I was using that I preferred being supervised by a specialist whatever I was doing. The invisible information which I had to pass to yeast was like treasure to protect.

PCR was consistently correlated with electrophoresis in order to check the result. Both tasks were repeated over and over again:

> Today, I had to do electrophoresis to check the result of last week's PCR. I tried my best to be as focused as possible. I started with agarose weighing. J. decided to watch me do it because last time I had added too much. When I put the flask with agarose into the microwave to dissolve it, J. tested me by asking what I should do next. I remembered that I had to add ethidium bromide before pouring the gel into the tray. I also remembered that I needed to put a 'comb' into the tray to obtain regular wells in the gel. When the gel settled, I pipetted the PCR product with a marker and the ladder to the first well. The last thing to be done was to connect the electrodes in a proper sequence. (5 December 2016) (Fig. 7.2)

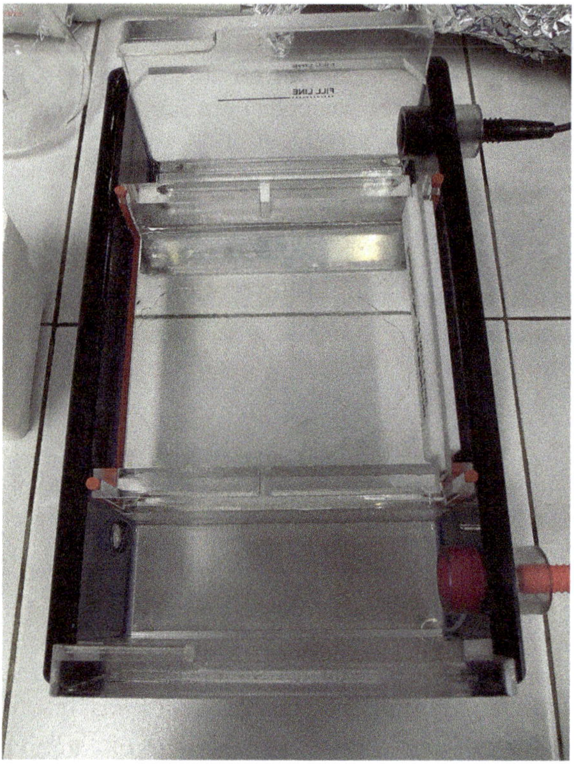

**Fig. 7.2** An electrophoresis device. Photo: Karolina Żyniewicz

All these operations were quite automatic. The most salient thing, my DNA, was a mere drop of liquid dissolved in the reaction mixture. So many procedures had to be performed around this tiny, hidden drop. The nature of my work changed when DNA moved between different 'living vessels,' such as bacteria (*E. coli*) and yeast (*S. cerevisiae*). Then, it was at least possible to watch the growth and general condition of the vessels. However, growth as such was not proof that the organism took what we had inserted into it. To make sure that this was the case, we needed to extract some genetic material from the vessel.

One of the fundamental questions I asked myself regarding the transmatterive production of *The Last Supper* was: Is DNA a liminal being, like cells are? Indeed, is it a being at all? Mackenzie describes her art&science operations on DNA with recourse to the coinage of 'lively matter':

> Lively material is an extension of 'living material' (…) to include the molecular biological material that is inherently vital to the process of life but that does not fall under any commonly accepted definition of life. I derive this term from my experiences embedded in the laboratory, where plasmids, viruses and DNA are not considered as life, yet when contained within the body of an organism, they act within the body and are thus lively. (2017: 22–3)

Transporting one of my genes from one living vessel to another was axial to *The Last Supper*. The protein coding part of the gene was not modified; however, a promotor, ADHp, was added to it to regulate the protein expression. A plasmid played the role of the transporter of the DNA fragment, and it was a lively vessel.

> Today we needed to isolate the plasmid to use it as a transporter of my gene to yeast. Then, we had to do digestion with special cutting enzymes, cut the plasmid in the proper places, and insert it into my gene. (10 July 2017) (Fig. 7.3)

When my gene was inserted into a plasmid, these two forms of lively matter became integrated. They were moved first to the bodies of bacteria and then to the bodies of yeast. This traveling from one body to another corresponded to the progression of the initiation ritual. Finding itself between different living beings, my DNA was definitely in a liminal position. It passed through a circular process (like the shape of a plasmid), starting from the human body—my body—going through the bodies of bacteria and yeast, and ending up in human bodies that consumed the food products containing my gene. I was also among the consumers of the final product, so my gene ultimately returned to the starting point. It was cognitively confusing whether I added something new to my organism or not by drinking the beer humanized by my gene. There was the same concern regarding the other consumers. Drinking the transgenic beer, they absorbed a human gene which they anyway already had. The subjectivity of the vessels was the crucial aspect of my project. The human gene inserted in yeast came from my body, a body with a particular identity of its own. The bodies of others could be perceived as a post-liminal

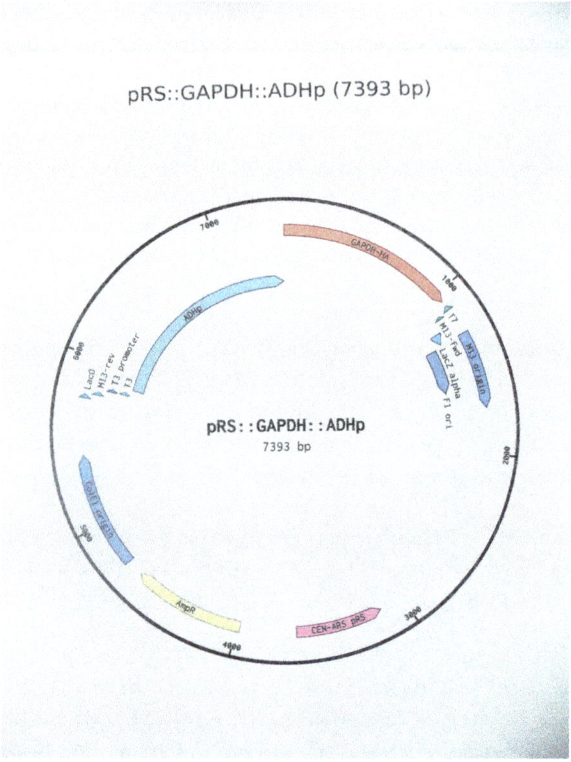


**Fig. 7.3** The map of a plasmid. Picture by Karolina Żyniewicz

destination where the lively matter—my gene—completed its travel. However, it would be possible to repeat the multi-vessel journey repeatedly by isolating the same gene from the consumers' bodies.

The idea of the body as merely a vessel for selfish genes was initially proposed by Richard Dawkins (1976), but his metaphor has since been used by others, including Mackenzie, who has employed it in her practice and writing:

> I must now insert my thought-as-DNA assembled in a lively plasmid vector, into a living organism. I cannot see the organism as a lifeless chassis in which to assemble inanimate parts. Instead, I choose to frame it as a vessel with all

> the bodily materiality this implies. Each step feels increasingly personal. I will do my best to relate it faithfully as I experience it. (2017: 124)

In her work, Mackenzie used bacteria (*E. coli*) as a vessel, transforming her thought-as-DNA (synthetic DNA). In my *The Last Supper*, the living vessels, *E. coli* and *S. cerevisiae*, were vehicles for a fragment of my genome. The operation of reverse transcription resulted in obtaining cDNA, that is, a direct copy of my original DNA.

Using bacteria in this project was very instrumental, as they were only tools, rather than being the research subjects.

> She explained that bacteria were transformative organisms, which meant that scientists could insert a protein into them. The bacteria would produce this protein at the expense of their own metabolism. (6 March 2018)

The notion that the organisms sacrificed themselves for scientific purposes sounded cruel to me. Additionally, the biologists said that the plasmid had to be shot into bacteria. It was a violent metaphor that stirred my imagination. I imagined the plasmid as a bullet, hitting the delicate bacterial body and destroying it. This image provided a meaningful bridge between *safe suicide* and *The Last Supper*, as both projects were based on a mixture of cruelty and care.

The actual process did not look very dramatic. We used a chemical method of transformation. It would have been much more drastic to conduct electroporation, in which bacteria are electrocuted by an electric impulse of 1.8–2.5 thousand V, or to use biolistics, in which cells are literally shot with DNA-covered beads. However, even in the chemical transformation that we applied, there was a moment of temperature shock forcing bacteria (in our case *E. coli*) to absorb DNA. For all this, the plasmid was not deadly to the bacteria, and they were supposed to store and multiply it. The destination point was yet another organism—yeast.

> The project is moving forward. Yesterday, we shot the plasmids into bacteria. Today, we need to isolate these plasmids to see if they have kept our insert. (13 March 2018)

Although plasmids are not lethal for bacteria, bacteria's bodies are destroyed when they are isolated.

The bacteria and their characteristics were not of central importance in the project. What they were supposed to do was simply to work as a living vessel. The laboratory workhorse, *E. coli*, was therefore an obvious choice. To select an organism to be humanized by my gene was equally easy. There was hardly any other option than the other laboratory workhorse, *S. cerevisiae*. Both organisms were readily available in the laboratory where *The Last Supper* was being developed, and they both had a long history of being used in laboratory research.

*Escherichia coli* (*E. coli*) has had a long laboratory career which dates back to its discovery 136 years ago. The main advantage of *E. coli* is that it is safe and easy to cultivate. The bacteria replicate at a very fast rate of 20 minutes per cycle. The four essential strains, which are commonly used in laboratories worldwide, have lost all the capacity to infect human intestines. Recently, Georg Church, together with Henry *Lee* and other scientists, has initiated research on replacing *E. coli* with *Vibrio natriegens*, bacteria that replicate even faster—every 10 minutes. Whatever their findings will be, it is bound to take some time before *Vibrio natriegens* replaces *E. coli*, given the long history of the latter's strong laboratory presence and use (Panko 2016).

As explained by Parapouli et al. (2020*)*, *S. cerevisiae* is a well-known model organism used in basic research on eukaryotic organisms. The yeast is unique thanks to its fermentation capacity and production of alcohol and $CO_2$. Its resilience to adverse conditions of osmolarity and low pH is also noteworthy. What sets *S. cerevisiae* apart from the other laboratory model organisms is its industrial applications. They are made possible by an aspect of the life of yeast which has been termed 'make–accumulate–consume.' This feature is based on a fascinating phenomenon called the Crabtree effect. Specifically, *S. cerevisiae*, even under aerobic conditions, does not use respiratory mechanisms to metabolize saccharides and promote biomass growth. Instead, the yeast produces ethanol and other two-carbon compounds via pyruvate. This system enables *S. cerevisiae* to accumulate ethanol, which is toxic to most other microbial species. The yeast can also use the collected ethanol for its growth. Importantly, laboratory strains are cultivated in very favourable conditions, and they would barely be able to survive outside the laboratory. It is an essential piece of information in the light of public fears that transgenic laboratory organisms can sneak out of the laboratory and contaminate 'nature.' I was surprised to learn that the yeast strains referred to as 'wild type' in the laboratory are not simply taken from the environment; they are prepared for further genetic operations. *S. cerevisiae* strains used for industrial purposes, mainly food and alcohol production, can survive in much harsher conditions than laboratory strains. This double function—laboratory and industrial—made *S. cerevisiae* the main protagonist of my *The Last Supper* project.

A transfer of genes to transformative organisms is like transplantation. When working on inserting the plasmid with my gene into bacteria, I had to consider a variety of outcomes. The bacteria could take in the plasmid with the insert, without the insert, or take nothing at all. That was why I needed to constantly repeat liquid and solid bacteria cultures and to check if they had accepted everything that we wanted them to have. In the optimal scenario, the transformative organism would grow with the designated DNA fragment inside and multiply it. To increase the chance of success, we decided to work with more than one strain simultaneously.

> J. came in, holding a Petri dish with new colonies. He explained that we could not ensure that every strain had the plasmid and the insert. That was why we had to use eight strains. I needed to add ampicillin to a new bacterium medium, doing everything close to the torch. Then I divided the medium between eight glass tubes, adding one colony to each. They went to the incubator. I was to isolate the plasmid from each of them on Monday. (14 July 2017)

It is difficult to count how many times I transformed bacteria by the plasmid to see later whether they had indeed been transformed. It was a long negotiation with a non-human agent (working as a vessel). The following stage of work—that is, inserting my gene into yeast—was much the same. We had to inspect more than one laboratory yeast strain to make sure that some of them would accept my gene. I did not observe the yeast under the microscope, but I learned how to photograph its solid cultures on the plates, using the camera to capture pictures of electrophoresis gel. In this way, I could be closer to these living vessels and recognize their colours and the colony's shape. However, taking pictures of the colonies did not mean carrying out precise microscope observations of what was going on inside the yeast organisms. The most important thing was to make the vessel hold the inserted lively matter and accept this burden as a part of its own structure.

The creation of transgenic yeast was not the last stage of transmattering in *The Last Supper*. The plan was for the genetically modified organism to run fermentation effectively enough to produce alcohol and bread dough. Even though *S. cerevisiae* is commonly used in baking and alcohol production, it was by no means obvious that the laboratory variants of the yeast would be as helpful in such undertakings as their industrial relatives. We in fact experimented with a food production strain which I had bought at a shop with ingredients for alcohol production, but it lost the plasmid. This made us realize that the humanized laboratory strain could not run fermentation properly. In the end, we mixed these two kinds of yeast, had the mixture lyophilized, and stored it for later. Thanks to this, we avoided repeating the process all along.

The idea of *The Last Supper* project was that my gene would be not only incorporated into the yeast body but also contained in the final foodstuff (mostly beer) produced in and through the metabolic processes of yeast. That was why the beer was not pasteurized or cleaned. We wanted the yeast to remain in the alcohol as the remnants of the organism that sacrificed itself for the project and the remainder of the prolonged and complex transmattering process. The yeast humanized by my gene did not develop features of human flesh (just like a tomato with a fish's gene does not taste like fish), but it was important that my genetic material had operated as a transmattering factor. The crucial point was that the remains of the humanized yeast made the beer dangerous and prohibited. The project had to comply with the legal regulations of GMO production and be moved outside the laboratory. That was an essential and exciting aspect of the project. Although I was not going to mass-produce humanized beer, I decided to apply to the General Sanitary Inspector for permission to hold dinner including the tasting of my humanized product outside the laboratory. The answer I obtained influenced the next critical stage of transmattering, when the project moved from the scientific context to the artistic one.

The legal regulations regarding GMO-based food are strict and complicated. In a sense, the law mirrors the general public's approach to laboratory-modified food. The regulations are also related to Bennett's observations about edible matter. The fact that what is eaten is incorporated into the eater's organism makes eating a very special activity. The idea that 'one is what one eats' is popularly shared, and this belief has resulted in the presumption that eating transgenic organisms makes one transgenic. This is, of course, not true; eating other organisms and digesting their genetic material does not change our human genome since the DNA one ingests is broken down in one's stomach.

As already mentioned, people react to GMO-based food the way they respond to the forms of matter defined by Kristeva as the abject. Transgenic matter seems to be 'outside the order.' In *The Last Supper*, the diners were asked to drink something contaminated by the essence of I-with. Some of them understood it almost as cannibalism, as eating/drinking a piece of another human being. Their response was not about the fear of consuming laboratory strains of yeast but about confusion related to absorbing the essence of life. As producers, we wanted the I-with's life essence to be present in the beer. The fact that the human essence was encapsulated in the remains of the yeast body was of relatively minor importance. Nobody from the audience analysed what the process of fermentation had been like. Explaining that aspect could potentially exacerbate the response of rejection. Knowing that the product contained something dead (dead yeast bodies) and contaminated by the human essence could make the storytelling traumatic.

## References


Blancke, Stefaan, Frank Van Breusegem, Geert De Jaeger, Johan Braeckman, and Marc Van Montagu. 2015. Fatal Attraction: The Intuitive Appeal of GMO Opposition. *Trends in Plant Science* 20 (7): 414–418. https://doi.org/10.1016/j.tplants.2015.03.011. Accessed 22 June 2022.

Dawkins, Richard. 1976. *The Selfish Gene*. Oxford: Oxford University Press.

Ferdman, A. Roberto. 2015. Why We're So Scared of GMOs, According to Someone Who Has Studied Them Since the Start. *The Washington Post*, July 6. https://www.washingtonpost.com/news/wonk/wp/2015/07/06/why-people-are-so-scared-of-gmos-according-to-someone-who-has-studied-the-fear-since-the-start/. Accessed 22 June 2022.

Mackenzie, Louise. 2017. *Evolution of the Subject. Synthetic Biology in Fine Art Practice*. Newcastle: University of Northumbria.

Panko, Ben. 2016. Step Aside, E. coli. Can a Faster-Growing Microbe Replace Lab Favorite? *Science* 352 (6293): 1503. https://www.science.org/doi/10.1126/science.352.6293.1503. Accessed 31 July 2022.

Parapouli, Maria, Anastasios Vasileiadis, Amalia-Sofia Afendra, and Efstathios Hatziloukas. 2020. Saccharomyces Cerevisiae and Its Industrial Applications. *AIMS Microbiology* 6 (1): 1–31. http://www.aimspress.com/article/10.3934/microbiol.2020001. Accessed 12 May 2022.

Robin, Michael. 2020. Why Do People Fear GMOs? *Ag-West Bio*. https://www.agwest.sk.ca/why-do-people-fear-gmos. Accessed 22 June 2022.
Venter, J. Craig. 2008. Synthetic Genomics: Designing Life. *Nature* 454 (7208): 554–557. https://doi.org/10.1038/454554a. Accessed 12 May 2022.
Watson, James D., Andrew Berry, and Kevin Davies. 2017. *DNA: The Story of the Genetic Revolution*. London: Penguin Random House.

CHAPTER 8

# Transmattering in the Art Gallery

This chapter is dedicated to the continuation of transmattering after a project leaves the laboratory to be presented at artistic institutions. This raises the question of whether exhibitions and artistic events should be regarded as a continuation of the laboratory trajectory or as an altogether new direction. Changing the context is a vital transmatterive factor, transforming language and meanings. The staff of art institutions and artistic audiences are new transmattering actors of the aesthetic/epistemic network.

Liminal projects based on biology are challenging to the institutions dedicated to exhibiting art as such settings have no codified rules for dealing with living non-human agents and biohazardous matter. Research facilities may have such protocols and be theoretically better prepared for exhibiting life, but they usually do so as part of their science-popularization agendas.

Putting life on display in a gallery requires several translations and mediations that work as an institutional adjustment. Once, when presenting my liminal activity, I heard that everything I had done in the biological lab when developing my liminal project was purely scientific, and that there was nothing artistic there. Consequently, I needed to switch to an artistic language that was not scientific for the person claiming that these two stages were disconnected. I could not entirely agree because the main idea of the project, its starting point that led to all the lab-based


K. Żyniewicz, *The Body as Matter in Art and Science*, https://doi.org/10.1007/978-3-032-13665-7_8

experiments, was artistic. What I could agree with was that objects shown at the exhibition were like a maquette of the project, or, more to the point perhaps, that this was a different theatre. It is impossible to share the transmatterive processes of the project production, learning, hands-on experiences, and casual laboratory talks. Even if these aspects are described in a text, this form of expression breeds different experiences than an art show does. The limited possibility of involving the audience in my creative experiences has always been a problem in my liminal practice. Each presentation of the two liminal projects described in this thesis was different, even though they were all about triggering the best possible experience and transmitting the most vital message.

So far, the first part of the *safe suicide* project has been exhibited four times, at the Contemporary Art Gallery in Opole, the Copernicus Science Centre in Warsaw, the Centre for Contemporary Art ŁAŹNIA in Gdańsk, and Kunstquartier Bethanien in Berlin. ŁAŹNIA was the only venue where my living/dying cells were actually present in the gallery. The project is dedicated to death, so I did not think it necessary to carry out the cell culture in gallery space, especially that arranging the sterile laboratory conditions at a gallery would have been very complicated and expensive.

I started developing the project with a clear idea of what would be shown when (if) it was actually put on display. Each presentation of the first part of *safe suicide* included round-shaped porcelain funeral pictures of the size of an average Petri dish. I had ordered them from a funeral company, asking them to put microscope pictures of my dying immortalized B lymphocytes on the porcelain, instead of the usual human faces (Fig. 8.1). The collection contains ten different images. Besides the photos, I also exhibited the remnants of cell culturing: cultivating flasks with dried-out cellular sediment on the bottom (Fig. 8.2), test tubes with a few millilitres of the medium, and cells stained with trypan blue (Figs. 8.3 and 8.4). My laboratory notebooks, research papers, and photo documentation formed another crucial component of the shows. Even before reading Kunst's book, I had in fact practised what she observed about contemporary art producing the documentation of processes, instead of artworks. The funeral pictures were also a form of documentation, but I perceived them as the most artistic element of my exhibitions. The remaining exhibits were transferred from the laboratory without any intervention. What I did was only build an installation of these objects as a material constellation telling a story. I expected the audience to connect the dots and construct the narrative for themselves. Narrative is a link between Kunst's

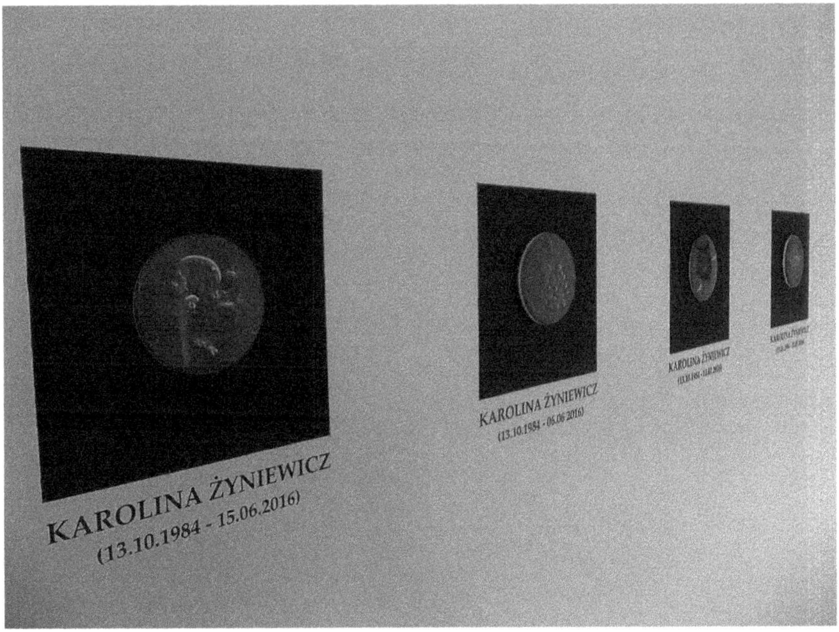


**Fig. 8.1** *safe suicide I*; funeral pictures of dying cells (*Non-Human Time* at the Contemporary Art Gallery, Opole, Poland). Photo: Karolina Żyniewicz

concept of contemporary art and Bardini's vision of it as encapsulated in his notion of junk art. Narrative is crucial in that it determines the value of artistic projects. The more skilled artists are in plotting stories the better they do in the sector of art production.

*Safe suicide* has never been presented as a solo exhibition; it has always been a part of a group show. This made the weaving of the story even more challenging because every time it had to be consistent with the story imagined by the curators. In a group show, a project is in dialogue with many other artistic products that have not necessarily been produced in the laboratory. The production processes are different, and the actors from the laboratory theatre are hired to perform in the gallery theatre.

The second stage of *safe suicide*, which was dedicated to cellular ageing, required different representations. I collected pictures from three kinds of staining: β-gal, BrdU, and 53BP1(DNA damage). The pictures were the primary material registering the progression of ageing. My artistic idea

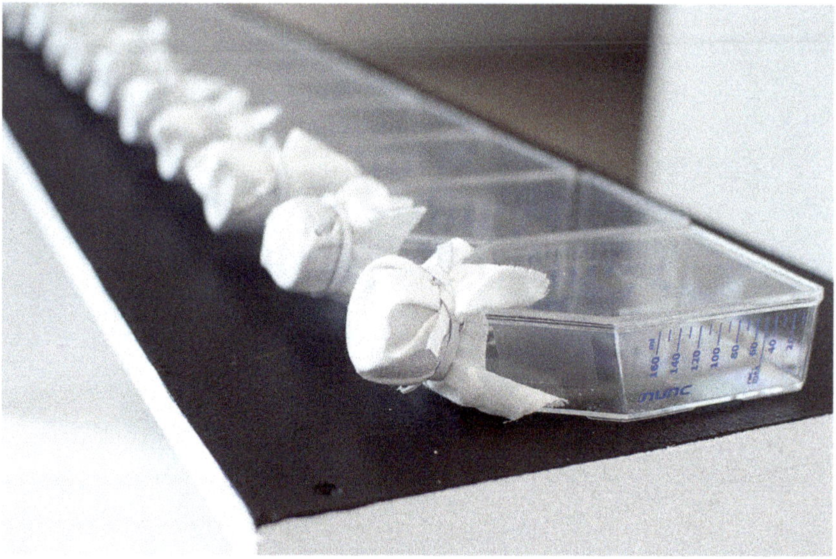

**Fig. 8.2** *safe suicide I*; cell culture flasks with cell sediment (*Non-Human Time* at the Contemporary Art Gallery, Opole, Poland). Photo: Karolina Żyniewicz

was to exhibit my photographic portraits accompanied by the pictures of the stained cells in order to show that ageing in *in vitro* was much faster than in *in vivo* conditions. This design was premised on playing with the notion of time as a transmatterive power. The images of cells demonstrated what had been happening with my fibroblasts for around a year. While the portraits of myself showed some changes in my face and hair, my skin did not appear to have become much older. The macro and micro photo documentation highlighted the temporality of my project. The viewers could see traces left at a particular time (Fig. 8.5).

So far, the second part of the *safe suicide* project has been exhibited twice, at the Labirynt Gallery in Lublin and at the Centre for Contemporary Art ŁAŹNIA in Gdańsk, where it was shown together with the first part (Figs. 8.6, 8.7, and 8.8). The video documentation of my work in two different laboratories was a crucial element of the show in Lublin. The video is over one hour long, so the audience usually saw only a piece of it (max. 20 min,). Integral to the presentation of the second part was also a display cabinet that contained: photo documentation of laboratory work;

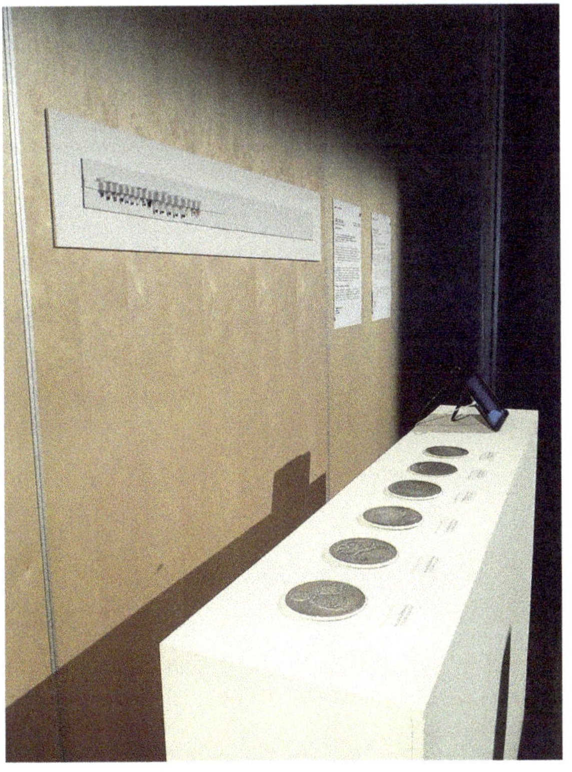

**Fig. 8.3** *safe suicide I*; funeral pictures of dying cells and tubes with cells stained with trypan blue (Przemiany Festival, *Temptation of Immortality*, Copernicus Science Centre). Photo: Agata Steifer

research results (tables, schemes); and a text that I read during working on the project. It operated as evidence of the long process of my laboratory work. The exhibition curators had put my part of the show in a separate box (Fig. 8.5), so the audience could feel they were entering a different world—a laboratory world. I have been told that this arrangement made people think that they were not experiencing an artistic work, but rather a scientific one.

Drawing a line between biology and biological art, both of which are based on working with life, can be interpreted as baseless. Kunst observes that:

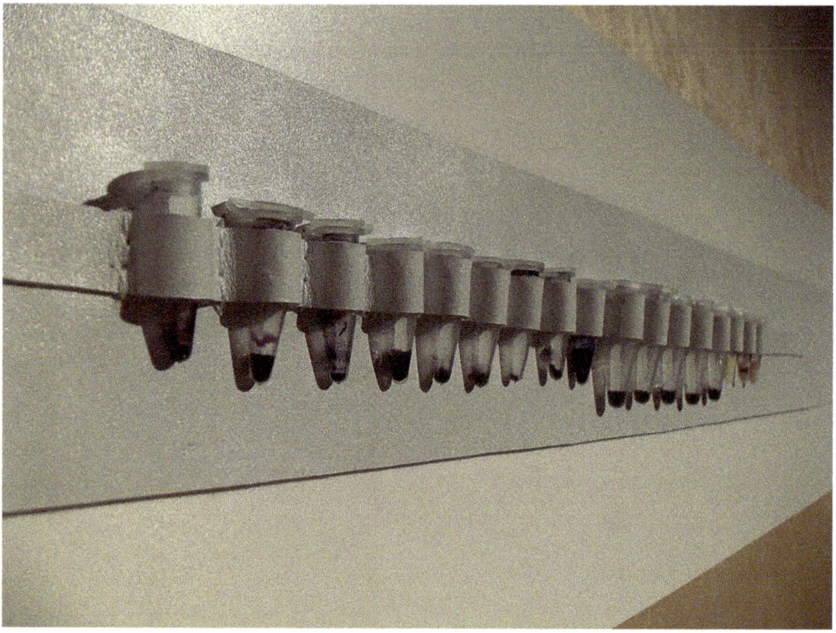

**Fig. 8.4** *safe suicide I*; test tubes with cells stained with trypan blue (Przemiany Festival, *Temptation of Immortality*, Copernicus Science Centre). Photo: Agata Steifer

> In contemporary artistic institutions, we actually do not witness artistic works, but the production and documentation of life as a pure activity by means of art. According to Groys, art has become bio-political in this sense; we are again confronted with the relationship between art and life, but in a totally different constellation. This relationship is 'characterized by the paradox of art in the guise of the art project, now also wanting to become life, instead of, say, simply reproducing life or furnishing it with art objects.' (2015: 173)

While this conception insists that contemporary art produces life, Latour's view is that science produces representations of life (1999: 30). Possibly, this is the main difference between art and science. Art seeks to generate an experience of life, demonstrate life, rather than providing an explanation of life, which is what science does. Art promises to offer an experience of real life, or of its documentation, in the fabricated capitalistic

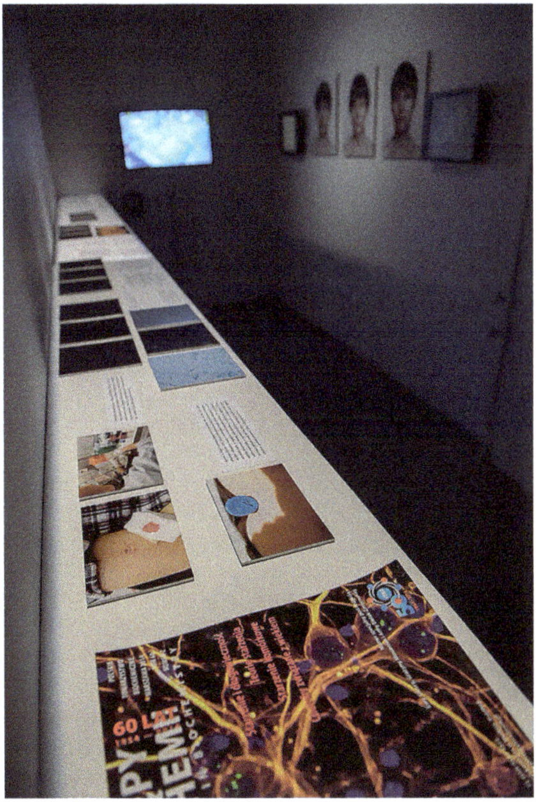

**Fig. 8.5** *safe suicide II* (the *Persistent Predilection* exhibition at the Labirynt Gallery, Lublin, Poland). Photo: Wojciech Pacewicz

reality. The problem is that documentation is a representation, so the experience as such is only available to artists. In the end, all sectors of the production of life tend to end up with and in representations, which give an illusion of something authentic.

Performative situations demonstrate life. They are authentic theatre pieces, and documentations are stage-sets. I happened to perform/demonstrate *safe suicide* once, at *Beyond Borders: Processed Body – Expanded Brain – Distributed Agency*, a show held at the Centre for Contemporary Art ŁAŹNIA in Gdańsk. My comments on the documentation above may suggest that the performative event should be much more successful at plotting the story. However, it was, in fact, not the case:

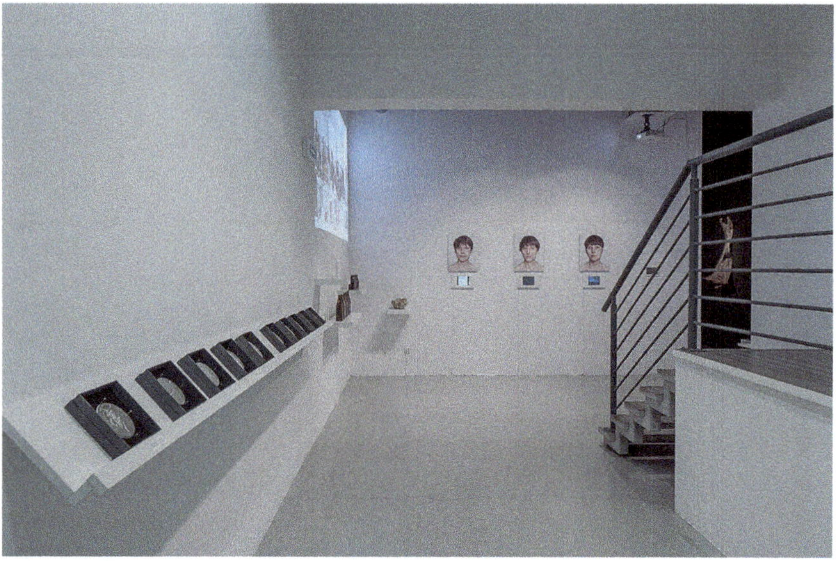

**Fig. 8.6** *safe suicide I and II*, shown together at the *Beyond Borders: Processed Body – Expanded Brain – Distributed Agency* show (Centre for Contemporary Art ŁAŹNIA, Gdańsk, Poland). Photo: Paweł Jóźwiak

I prepared the performance room, trying to make it evoke the atmosphere of a chapel. The room had black walls and floors; the windows were covered in black cloth. A black altar was the main object in the room, where I would add hypochlorite to my cells to show their dying to the audience. I had a special black dress on and a black box with all the accessories with me. The stress level was rising by the minute as the beginning of the celebration was approaching. Everything could go wrong, from problems with cells to the malfunctioning of devices. I had created a precise scenario in my imagination, planning about 30 minutes of action. The biologist helping me work with cells in Gdańsk had given me a fresh bottle with cells, a new multi-well plate, and a Falcon tube with hypochlorite. I put everything into the black box and entered the room, walking across it directly to the black altar. I set the laboratory timer to count the time, letting me know when I should finish my action. However, extremely nervous and excited as I was, I forgot to set the timer directly after coming to the black table. I only did so 10 minutes later. That made me even more stressed out, because the plan I had prepared before was ruined. The room was full of people, and I was doing my best to focus on my action, not on the audience.

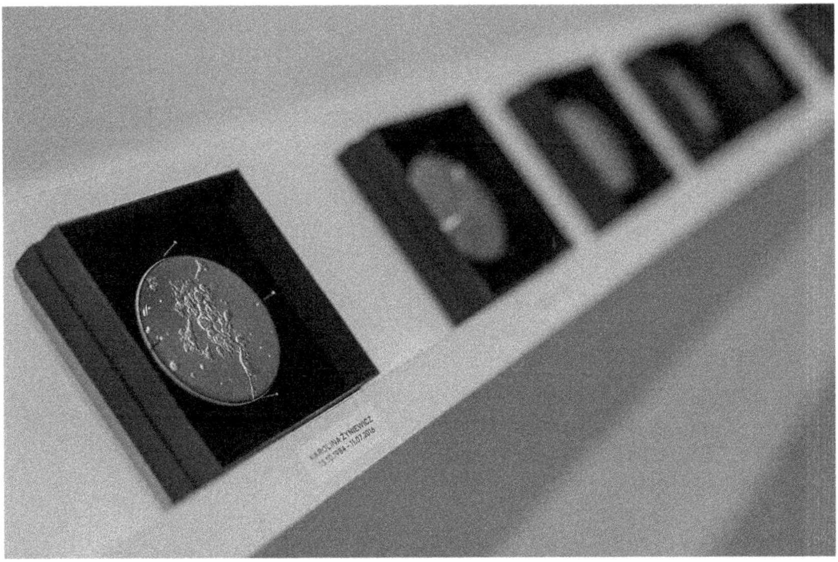

**Fig. 8.7** *safe suicide I*; funeral pictures of dying cells (*Beyond Borders: Processed Body – Expanded Brain – Distributed Agency*, Centre for Contemporary Art ŁAŹNIA, Gdańsk, Poland). Photo: Paweł Jóźwiak

> I had to pipette 0.5 millilitre of the medium with cells to the plate wells and add different volumes of hypochlorite to each well. I had checked in the laboratory how fast hypochlorite worked, and I knew that with the highest concentration, around 0.5 millilitre, the cell reaction was almost immediate. I did not take it into account, though, that the bottle had been shaken during the transport from the laboratory. B lymphocytes are non-adherent cells, so they do not attach to the bottom of the cultivating flask. Pipetting made them move even more. That made finding their picture under the microscope very tricky. I entirely forgot about my plan, so I had to improvise. (a note taken one day after the performance) (Figs. 8.9 and 8.10)

The fieldwork note shows that the performance was based on the direct transfer of laboratory activities to the gallery. The only difference was that the gallery iteration had a pre-designed scenario. I added a volume of hypochlorite (checked before) to my cells to cause apoptosis. Nothing more spectacular beyond the simple laboratory activity was happening. The project documentation was displayed in another gallery room. As I

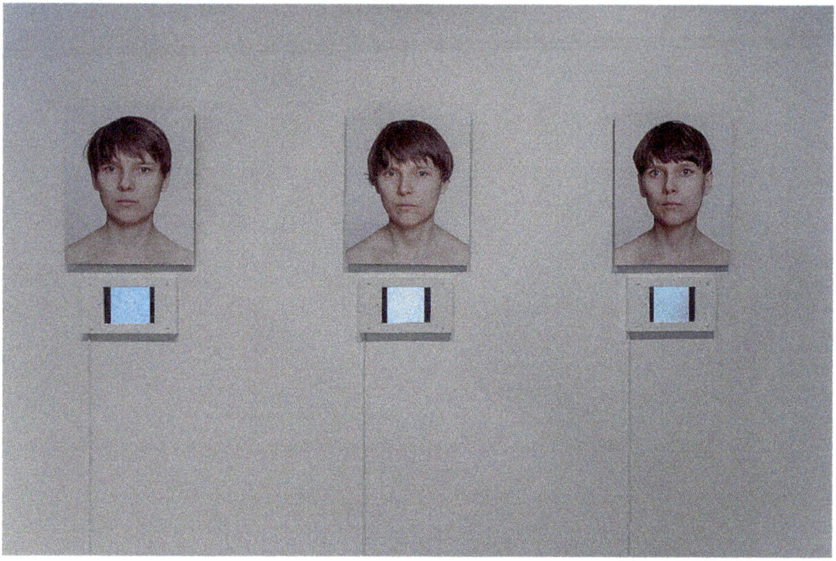

**Fig. 8.8** *safe suicide II*; regular and cellular ageing (*Beyond Borders: Processed Body – Expanded Brain – Distributed Agency*, Centre for Contemporary Art ŁAŹNIA, Gdańsk, Poland). Photo: Paweł Jóźwiak

found out, the action itself had been neither attractive to the audience nor intelligible to them without a textual description. They had failed to realize that I was demonstrating life and death. This situation made me realize that an unexplained action could not be adequately experienced. It appears that the life enacted in the gallery theatre must be completed by a text or documentation. Experiencing is impossible without understanding, especially in the case of performances that mimic the laboratory theatre. The product is defective.

A further question provoked by contemporary art as the production of production for the capitalistic market concerns selling. When is the project actually sold? Is it sold when it is accepted to be exhibited by an institution, or does it have to be bought by the audience? In these terms, it is difficult to judge whether I successfully sold my laboratory trajectory. While the gallery did pay me money to exhibit the project, I am not exactly satisfied with its reception by the audience. The long transmatterive journey, so crucial for I-with, has been bought by an institution, but the

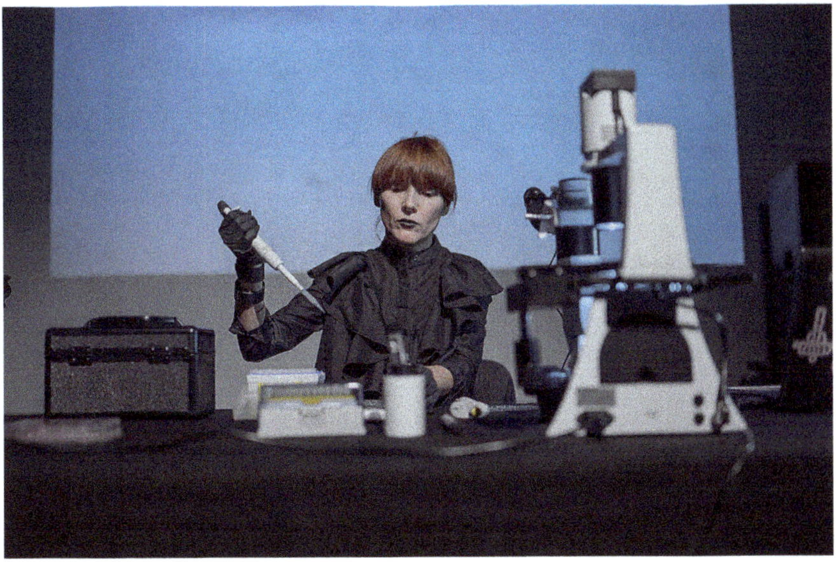

**Fig. 8.9** *safe suicide ceremony* (Centre for Contemporary Art ŁAŹNIA, Gdańsk, Poland). Photo: Paweł Jóźwiak

audience have only taken the visual trappings of my play without grasping its core. This selling transaction worked as a crucial transmatterive power that questioned the entire transmatterive laboratory process.

*The Last Supper* project has only been exhibited once. This took place as part of *Beyond Borders: Processed Body – Extended Brain – Distributed Agency* in Gdańsk. This was preceded by the most critical performative event which had to be arranged in the laboratory area because, as already mentioned, I had not obtained the official permit to organize an event involving the consumption of Genetically Modified Organism (GMO) products outside a research facility. That meant that I had to arrange the event in a laboratory classroom. It was an on-camera performance. My idea for the event was to reproduce Leonardo da Vinci's famous painting *The Last Supper*. Twelve selected visitors were fully incorporated in the performative situation and cast in the roles of the apostles. They drank beer containing the remains of yeast modified by my gene, and they ate bread the baking of which was made possible by the death of the transgenic organism. The eating-and-drinking was a shared experience that was

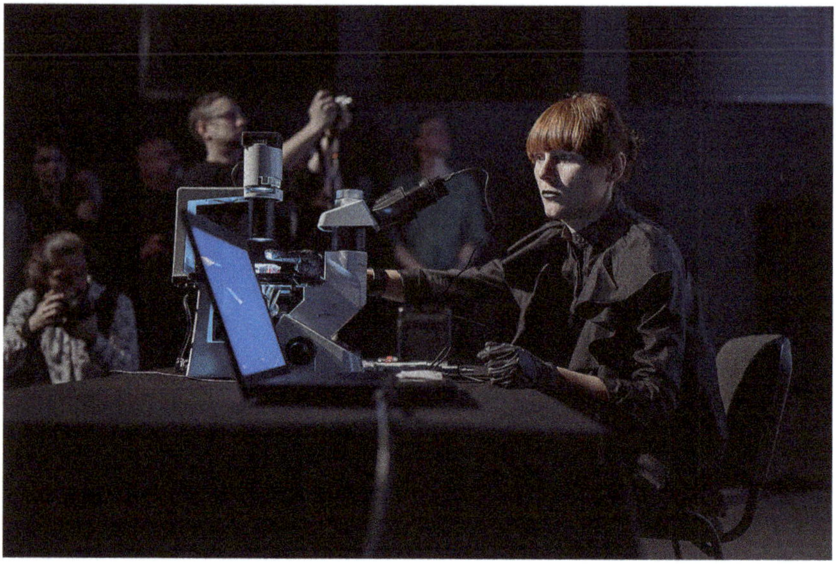

**Fig. 8.10** *safe suicide ceremony* (Centre for Contemporary Art ŁAŹNIA, Gdańsk, Poland). Photo: Paweł Jóźwiak

discussed afterwards. The discussion also addressed the scientific process of the development of the project, the meanings of the project, and the ways in which the audience understood it. The event was a lively situation, but it has only survived as a representation. My scientific collaborator and I had to produce beer and bread again for the *Beyond Borders* exhibition. However, this time the audience were not allowed to consume the products. They became museum exhibits. The bottles remained closed (Figs. 8.11 and 8.12).

A gallery worker guarded the container with the bread and stopped people from tasting it. The video recording of the first tasting was screened large-scale on one of the white cube walls. It had been filmed without sound, so it looked like a dynamized fresco. Like *safe suicide*, *The Last Supper* was re-presented and meaningfully transmattered. Both transmatterive episodes were different. In my view, only the first performance had a ritualistic dimension and demonstrated life, including everybody in the action. Watching the performance representations as such could not be as engaging. The official regulations on the circulation of GMO outside the

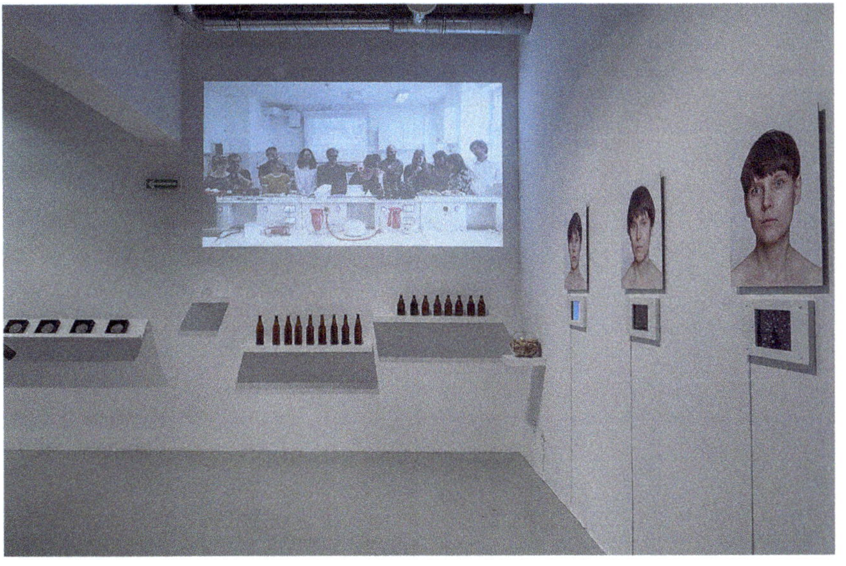

**Fig. 8.11** *The Last Supper* (*Beyond Borders: Processed Body – Expanded Brain – Distributed Agency*, Centre for Contemporary Art ŁAŹNIA, Gdańsk, Poland). Photo: Paweł Jóźwiak

lab forced me to transmatter the previous script into a non-interactive setup. The transgenic yeast has been stored and may be used when organizing a performative dinner will again be possible.

In *Give Me a Laboratory and I Will Move the World* Latour describes an exciting phenomenon of the laboratorization of social life, which, as he claims, helped Pasteur to resolve an agricultural problem in laboratory conditions. Pasteur had to extend the laboratory conditions onto social space outside the laboratory. Put differently, reality had to be laboratorized. With regard to the presentation and representation of liminal projects, I propose that liminal projects based on laboratory work are a form of laboratorization, but its goals differ from those pursued by Pasteur. He worked to solve a defined problem and had to translate micro-scale processes into macro-scale ones. In simple terms, he transferred knowledge. Liminal projects share knowledge as well, but they do not aspire to solve any problem. A problem may be solved in the process, but it is never the primary purpose. Through liminal projects, society learn about liminal

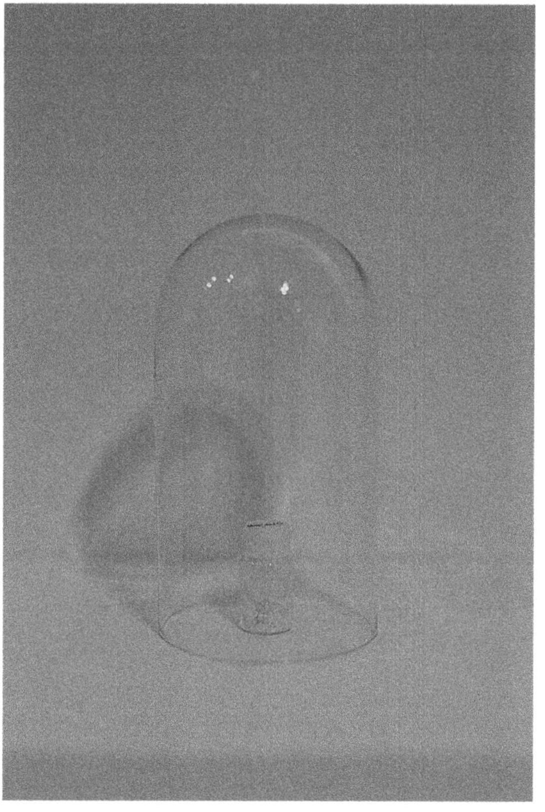

**Fig. 8.12** *The Last Supper*; lyophilized transgenic yeast (*Beyond Borders: Processed Body – Expanded Brain – Distributed Agency*, Centre for Contemporary Art ŁAŹNIA, Gdańsk, Poland). Photo: Paweł Jóźwiak

lives, which supports the general becoming-liminal (to use Squier's terminology). An artistic practice called bioart transfers biological knowledge to the art world to foster reflection on it. Potentially, exhibiting institutions are more easily accessible to common citizens than traditional scientific institutions and sterile workplaces. However, both fields—contemporary science and contemporary art alike—are perceived as elite genres. Artists translate scientific research into the matter of reflection and critical discussion. As a result, scientists and all other people can deliberate on cutting-edge scientific methods and discoveries. This highlights the fact that science is of interest not only to scientists since it affects everybody's reality. Multiple liminal projects transfer not only scientific topics but also living organisms and laboratory equipment to exhibiting institutions. This aesthetics produces a peculiar impression that the laboratory is ubiquitous, while such living organisms prove that culture and art are not the human monopoly. As liminal practitioners are interested in the life sciences and follow the humanities discourse, the laboratory is becoming expanded and augmented. Owing to their emotional component, liminal practices capture the audience's attention much more readily than scientific papers and lectures do. However, artistic projects based on science often appear too complicated for the audiences, who are accustomed to more traditional art. Even if liminal practice transfers knowledge, it should not be confounded with the science popularization or advertising.

Besides exhibiting projects, there is another way of sharing the daily laboratory work with a bigger group of recipients. This way is provided by social media. Most liminal practitioners have accounts on Facebook, Instagram, and/or Twitter. They use these tools to post information, images, and stories showing the progress of their work. Such posts include photos of laboratory equipment, microscope pictures, and even selfies of liminal practitioners sporting laboratory coats and surrounded by their scientific collaborators. This daily sharing shows that the projects are indeed a reality of these individuals. Leaving traces and providing evidence that the creative process is unfolding in the laboratory and is based on scientific methods are other functions of liminal practitioners' social-media presence. This bolsters the credibility of the laboratory trajectory. By presenting the work process in this way, artists can also show elements that cannot be moved to the artistic context because of the legal regulations in place. When exhibiting and presenting *The Last Supper* at conferences, I often heard that investing so much time and work in humanizing the yeast to produce GMO products did not make any sense. The audiences suggested that I should serve a regular brand of beer accompanied by my story of transformation as an artist. This, however, has never been what my work was all about. I am always focused on my credibility and professional ethics. I do not want to lie. The transmattering of a project by transferring it from science to the field of art should not be based on illusion and fiction, at least not without a sound explanation of that.

The movement from producing liminal projects to their presentation/re-presentation, described above, encourages revisiting the feminist understanding of the material and recalling Mitchell's categories of biological art: prophylactic and vitalist bioart.

As already mentioned, all presentations of the *safe suicide* project, except one, were based on photographic objects and remnants of my laboratory work. The process was materially re-presented. A relevant question is thus whether presentations of a liminal project are derivative re-presentations—namely, re-presentations of re-presentations produced by science. If prophylactic bioart is not based on hands-on laboratory experiences but is rather loosely inspired by laboratory aesthetics or specific biotechnology problems, its final objects cannot be regarded as second-hand re-presentations. It is even unwarranted to call the makers of such works liminal practitioners. The situation is meaningfully different if a person works in a biological laboratory and carries out regular research but for one reason or another decides, like I did, that the presentation does not require having a living being in the exhibition space. This, in my view, makes for a translative or transmatterive strategy.

In terms of the Science and Technology Studies (STS) research and the New-Materiality approach, the prophylactic tactic does not meet the requirements of 'being many', knowing by performing, and life demonstrating. Characteristic of scientific posters, slide shows and conference presentations, microscope pictures, schemes, visual data, and audio recordings displayed in a gallery are complemented with something 'artistic,' in the sense that they are adjusted to the language of art. Scientific representations are re-presented to be readable in the artistic context.

Vitalist bioart, the other bioart tactic identified by Mitchell, appears much closer to performative visions of the material world than the prophylactic type. Barad has called for stepping beyond representation-based epistemology. She has developed the notion of onto-epistemology, which encompasses knowing by existing/performing. In the light of this concept, representationalism is artificial and redundant. It is enough to be consciously re-configured and to 'be many,' or, as Tsing would put it, contaminated by entanglement (2015).

Mitchell's definition of vitalist bioart suggests that this is a strategy of practising/demonstrating life. Given this, exhibitions are not about presenting or re-presenting; rather, they are about creating a performative and relational situation. However, similarly to the definition of the prophylactic tactic, the notion of vitalist bioart is neither obvious nor straightforward. If living agents are present at the exhibition, the audience have an opportunity to interact with them or at least observe them living. However, even these multi-species meetings happen as part of a particular scenario, or rather, of a performance mise-en-scène. To make a relational situation happen, artists must make it attractive. There are containers designed for displaying living agents, and there are rules for interacting with them. These containers and boxes work as vessels for vessels (in the light of my remarks on organisms as vessels for lively matter).

At the same time, an aesthetically attractive manner of presenting certain living beings and liminal lives is not enough. There must be a story to tell, a question to ask, a discussion starter. Without such an element, the exhibits remain educational objects, rather than being liminal projects.

My two liminal projects that lie at the core of this text indicate that two categories barely suffice to classify liminal projects. Their existence is rather fluid. All exhibitions or other events spark new possibilities and come with different on-site requirements. The *safe suicide* project has been re-presented a few times, and a performative situation has also been staged with participation of my living/dying cells. *The Last Supper* project involved other difficulties; it was first put on as a performative 'being together,' which was then re-presented in an installation based on material evidence. Which fashion was more suitable and/or effective is difficult to judge without considering the audience. I believe that for a liminal project to be successful, the audience must be incorporated into its network, whereby the central question is: What can be regarded as a credible indicator of such an incorporation? Is it any response from the audience, or is it perhaps understanding? When can a liminal practitioner feel that they did well?

When depicting my autoethnographic work, I mentioned that my initial research aim was to examine the audience of my liminal projects. I always found it crucial to know whether my liminal practice worked and, if so, how. Later, I realized that, emotionally, this was very complicated to me. I wanted to sell my art so that the audience understood my intentions, appreciated them, and were intellectually and emotionally moved. Meanwhile, people either failed to fully grasp what I wanted to communicate through my project or my exhibition or complained that it was too scientific. That made me wonder about my own expectations concerning the recipients of my creative messages and their position in my projects. Was I able to incorporate the audience into my plan? Could I transmatter them as I transmattered myself and my biological pieces? The answer I finally arrived at was that it was not really the case, and for several reasons, too.

Let us first consider the prophylactic kind of bioart. The audience only explore what the artist has decided to select from a more extensive collection of available representations. The main inspiration or reason behind creating can be explicated in a description displayed in the exhibition space, but most of the audience are not likely to read it. Consequently, their reception remains superficial and boils down to 'this is nice or not nice.' A lot of viewers—and some artists—still believe that art should talk by itself without being supported by textual commentaries. This was what keenly bothered me whenever *safe suicide* was statically represented. I was as a rule present in the exhibition space for some time to explain the project to the viewers. They were not always happy that the artist was there and listened to what they said about the exhibition. I had no patience to listen to the visitors saying: 'such nice pictures' or 'are these bacteria or what?' All the people with whom I talked admitted that the project would not be understandable and even moving without my explanation. They needed to listen to the transmattering story. This implied that I produced what Bardini dubbed junk art. The material details on display were vacuous without the artist's comment. From my perspective, the most critical issue was that the audience did not grasp what had happened in the laboratory. The visitors did not understand that I had transmattered my own body, that I had been exploring the theme of suicidal death, and that I had been working with my own tissues. While pictures of blood-taking and my skin biopsy were displayed in the part of the show dedicated to the research process, that component tended to be overlooked. The viewers very rarely connected the dots in the way I wanted them to. Even if the shortcomings of the representation might have been blamed for this limited connection, the fact is that the performative event during which I killed my cells in front of the audience did not change much.

> The overwhelming feeling was that the audience were bored because I struggled to get a picture projected on the wall behind me. At that very moment, I caught a glimpse of people leaving the room. I eventually managed to catch the picture of dying cells a few times, but I moved the plate too fast. To me it was obvious that my cells were dying, but I felt I was the only person to know that. I heard one member of the audience saying: ah, this is playing with bacteria, nice. That made me angry and sad because I took this as an indication that the people did not understand what I was doing. Thinking about it at the moment was not a good idea. Because I had set the timer 10 minutes into the event, the performance lasted longer than I had planned. I packed everything into the black box again and left the room, feeling that the performance had been a disaster. We had decided with the organizers not to add a separate description of the performance, seeing that it was a part of the project presented at the exhibition. We had assumed that people would get the relation. They did not.

> There is always a risk that people will not read the description. I wondered if hanging a performance description on the door to the room would have changed something. The biologists among the audience tried to comfort me by saying that the experiment had gone well. I caused apoptosis, and it was visible on the big screen. I realized that they were actually the only people to make sense of the performance. People from the art world suggested that I make the action more theatrical, using text or maybe sound, but to me it would be a decoration, something artificial. I wanted to make it honest and straightforward. This experience made me wonder if it made sense to put so much effort into laboratory work and then translate it into the language of art if nobody understood and cared. (19 January 2020)

Both performative situations—the *safe suicide* ceremony and *The Last Supper* celebration—were supposed to be provocative and push the audience out of their comfort zone. I used my usual trick, which Kunst made me realize was a common one in contemporary art:

> Profanation is a powerful procedure because it neutralizes what it profanes; it takes the aura away from things and people. Profanation is a highly important procedure in twentieth-century art and is deeply inscribed in the paradoxical relationship between art and life. Art is thereby established as a sort of field of radical events, a field of the potentiality of rebelling against the rigid structures of contemporary life; art also enables the autonomy of the artistic object. (2015: 35)

Safe profanation is an aspect that makes contemporary art a perfect limivoid experience. The art scene is a theatre where tragedies can be securely staged. Extreme emotions can be safely experienced without incurring any harmful consequences. In the *safe suicide ceremony*, a problem appeared when it came to the audience experiencing profanation. First, I profaned the gallery space by bringing in the biohazardous laboratory matter of my living cells. In the following, even more important, act of profanation, I conducted a ritualistic killing of my cells. However, as far as I know from immediate reactions and from what I heard later, the audience did not realize that a profanation was happening. The narrative element of junk art was missing.

The role of the exhibiting institution was crucial in this respect. As Kunst insists, "The museum is (…) the space where profane artistic procedures are isolated and given an almost ritualistic character, again returning to the field of the sacred rather than the common" (2015: 37). The staff of the gallery where the *safe suicide ceremony* took place did their best to frame the event in the most effective way. Generally, projects with a laboratory past are challenging to art institution workers because they are not familiar with the laboratory trajectories experienced by liminal producers. In my case, a Gdańsk-based laboratory was helping me to prepare my cells for the performance. The institution supplied all the necessary devices, such as a microscope, a microscope camera, and an overhead projector. Everybody was helpful and receptive to all my requests. Despite these efforts and goodwill, things did not work the way they should have. This makes it clear that, like the laboratory trajectory, the gallery trajectory depends on how a network hangs together and operates. The way in which all its actors collaborate determines the value of the experience. During the *safe suicide ceremony*, the audience (except biologists) was the element that did not 'click.' Rather than being a reason to blame anybody, this is a motive to try again. The audience of contemporary art is a group of 'emancipated spectators,' to use Rancière's coinage (Kunst 2015: 60). They need to make an effort to participate in the production of production and to join artists and other art workers in being responsive and co-responsible.

The experience described above proves that it is challenging to continue transmattering outside the laboratory. It looks like the meaning of the liminal project, all its stages, and elements, was unexpectedly deformed. The distance between the starting phase of transmattering (having my blood taken) and the scene of the gallery events seems to be very long. On the whole, actors playing the leading roles at the stage of laboratory work are not directly related to those performing on the art-world scene. My own body and I-with are the only link that recurs throughout the chain of transmattering stages. It does not mean that biologists are not active when the projects they have co-created are exhibited. Their role is transmattered into being co-artist. They are usually surprised by the transformation. Unlike myself, they cannot include this artistic achievement in their professional curriculum vitae. In my liminal projects, the biologists who had helped me in the laboratory were usually mentioned as 'scientific support.' As the account of my performance shows, scientists can also play the role of, so to speak, enlightening the audience since they understand the scientific technicalities of liminal projects. They are also critical when a project simplifies scientific knowledge to create a particular aesthetic impression. Scientists can see in liminal projects what is usually overlooked by the general audience. Nevertheless, to conclude that liminal projects are only intended for scientists would be an exaggeration. As already mentioned, liminal projects are not developed as didactic models. Their role is to spin a broader cultural narrative around science and technology. However, to reflect on something that one does not understand is awkward, if not entirely unviable. The audience should grasp the main point and the liminal form of making this point. This failed to happen in my performance of the *safe suicide ceremony*. What should be done to clarify the project/action and to spark deeper reflection at a defined moment? First of all, the project description drawn up by the artist must be transformed, usually shortened, to make it effective in promoting the event. That can meaningfully change the main idea.

Another thing is that not all the members of the public carefully read event descriptions before attending. As a result, the meaning may remain impenetrable to visitors. Admittedly, even without comprehending the main idea, every viewer can interpret the exhibition/performance for themselves, but such interpretations can veer far from the artist's fundamental concept. While art has always been appreciated for giving the audience the freedom to understand it in their own ways, junk art is different; it has to be understood correctly. In retrospect, the *safe suicide* performance did not work because my action was too modest; it was not engaging or entertaining enough, as indicated by what I heard from the curator, the gallery staff, and other artists. What I did was limited to transferring a laboratory activity to the gallery space and creating a sacral scenario. My reasoning was that because it was about ritual killing and the general dynamic of death, the performance should be calm and simple. As such, I agree that looking at a person pipetting and a biological image screened on the wall was boring for the audience who did not know what the meaning of my action was. Those who did not understand what was going on could not possibly be transmattered by the action.

The performance linked to my *safe suicide* project has amply shown the significance of the relation between artists, art institutions, and the audience. *The Last Supper* celebration was exceptional because no artistic institution was involved, and the audience were invited to experience an artistic performance at a scientific venue. The event was mainly announced on social media. My plan was to reproduce da Vinci's *The Last Supper* scenario in a didactic laboratory and to video-record the moment of the consumption of my GMO products. Unlike in the performance of the *safe suicide ceremony*, the area of science in this performative event was profaned by an artistic activity and the presence of the audience expecting to experience something that usually happens at art galleries. The room in which the action was put on could contain very few people. The front of the room was a stage, and the camera was placed at the back. In my social-media announcement, I asked people to play the roles of the apostles. The rest of the audience could be present behind the camera. The event venue was not conspicuous; the laboratory was located on the second floor of a many-winged building, one of many on a spacious campus. Many visitors lost their way and came delayed. People playing the roles of the apostles became the main characters of the artwork. They had agreed to ingest the GMO product, to be recorded while doing that, and to have their image shown to the audiences of later exhibitions. There was something ritualistic about the performance, but the ritualistic framing was not designed by an art institution. I created the *communitas* of eating and drinking together, which was not allowed anywhere else and besides that one-time occurrence.

The apostles I had assembled had to trust me as a ceremony leader and also me and my scientific collaborator as the producers of the beer and the bread that we had correctly followed recipes when making them. My gene, a piece of my body, had been absorbed by other bodies. All the participants in the event knew that behind this simple act of eating and drinking, there was a symbolic meaning and a provocation to discuss social controversies around GMOs. The discussion took place immediately after the degustation, and I had time to listen to people's descriptions of what they had felt during the symbolic consumption.

We were together in that. When the recording ended, I offered the GMO products to the other visitors, not the apostles. Some of them refused to try the foods, and the two main reasons they cited were disgust (an association with cannibalism) and doubts about whether the beer had been produced correctly. Most of visitors who had consumed the products said that they tasted excellent.

Some of them took more than one serving. Because the event was held in the laboratory, many biologists were present. It was much easier for them to attend the event at their home institution than at a gallery. Many of them had been able to witness how the project had come into being. The performative event was based on the activities that they knew well. The gathering can be regarded as an empirical confirmation of 'being many'; it was about being and performing together. Intriguingly, the event happening at a non-artistic venue worked much better than the one held in a very prominent gallery. Perhaps the scientific context provided the audience with a more precise explanation of the action than any description. Although no art institution paid for the performance, it was symbolically bought, as a value, by the audience. The experiment with mixing the contexts proved that art could play life wherever. Putting an art label on things sufficed.

The exhibition of representations of the performative supper was incomparably less powerful. The exhibition audience were unable to feel involved in the ritual, mainly because consumption was impossible. There was no room for discussion and experience-sharing. Transmattering went out of my control the way it had done in the *safe suicide ceremony*. *The Last Supper* performance had transmattered I-with much better.

Representation, mediation, and re-mediation are relevant and helpful concepts in making sense of transmattering as it happened in the laboratory and in the gallery. Thacker emphatically regards biomedia:

> not simply as a technological instrumentalization of the (natural, biological) body, but as a unique domain in which biology is a technology (in fact, a better technology than any we can build). It is not just that the medium is the message, but that biology is the new medium: the medium is a message, and that message is a molecule. (2004: 48)

Taking into account that biology is a medium, Thacker also asks if the body in and of itself is a medium as well. Thacker's question makes perfect sense when we think of the body as a tool of expression and communication. Biology and its methods can re-mediate the body, translate it into the language of science, and re-present it on the social scene. Thacker adopts and champions Jay Bolter and Richard Grusin's definition of re-mediation: "'Remediation' is a concept that describes the ways in which any historically situated medium always re-mediates prior media, and thus also re-mediates prior social and cultural modes of communication" (2004: 55).

The body can be construed as the starting point of the re-mediation chain; it is a medium that is first re-mediated by biology and then by biomedia, new technologies transforming the body into laboratory liminal lives and, subsequently, into digital data. However, the process is not one-directional, and it can unfold in the opposite direction: "the biological 'informs' the digital, just as the digital 'corporealizes' the biological" (Thacker 2004: 54). This complexity of mediation and re-mediation phenomena is remarkably consistent with Barad's idea of constant re-configuration. How are liminal practices situated vis-à-vis these processes?

Do they simply re-mediate biology? What implications does this have for exhibiting, as described above? Answering these questions is facilitated by Sarah Kember and Żylińska's explorations of mediation and re-mediation. In their definition:

> Mediation does not serve as a translational or transparent layer or intermediary between independently existing entities (say, between the producer and consumer of a film or TV program). It is a complex and hybrid process that is simultaneously economic, social, cultural, psychological, and technical. Mediation, we suggest, is all-encompassing and 0indivisible. This is why 'we' have never been separated from mediation. (2015: xv)

The definition points out that everything that we do is mediation. The process is much more complex than translation, and it is not only about producing representations. Like Thacker, Kember and Żylińska make re-mediation part of their examination. They state that re-mediation is a 'mediation of mediation' and every instance of mediation is re-mediation at the same time.

Crafting a message, producing representations, and performing are forms of mediation and re-mediation. In both my projects, the body was a starting point. It was re-mediated by biology, then by art, and finally by the audience's bodies. Some of the important questions this involved are: When do mediation and re-mediation work well? Does transmattering depend on mediation and re-mediation? As already explained, from my perspective, the mediating/re-mediating process did not go well in exhibiting/performing *safe suicide*. However, from the perspective of some visitors, it was not a failure. As far as the transmattering of I-with is concerned, mistakes in mediation and re-mediation disturbed the gallery transmattering trajectory. The performative event of *The Last Supper* in my view excelled the presentation of the project as a set of exhibits. Mediation turns out to be always different for every mediated and re-mediated subject. It cannot be fully controlled. Kember and Żylińska's perspective, which stresses that these processes are hybridized, is comforting to liminal practitioners like myself. We can say that we had been doing whatever could be done until we lost control.

Mitchell's distinction between prosthetic bioart and vitalist bioart can give the impression that vitalist bioart brings the audience directly in touch with living beings participating in the liminal project. Living exhibitions foster a form of coexistence with non-human actors. To a certain extent this is indeed the case, but flasks and devices keeping these non-human actors alive must not be overlooked or forgotten. Mackenzie brings this aspect into relief by talking about 'behind-glass' aesthetics resulting from:

> the tendency (often a necessity) to house works behind glass, plastic or other forms of protective covering. I use 'behind glass' to describe the barrier between the audience and the work, but also to allude to the expressly scientific nature of the barrier in vitalist bioart. Artworks are often live and/or fragile and there may be concerns around contamination (of the audience or of the work itself). (2017: 34)

The mediated mode of contact is often transferred to the artistic context, with this solution informed not only by artists' fascination with laboratory aesthetics. As Mackenzie stresses, it is imperative to maintain sterility when things considered a biohazard are placed in a gallery. Mackenzie also points out that this springs from the care about both sites: the living actors of exhibitions and their human visitors. It is a fascinating phenomenon that can be regarded as an illustration of cutting together-apart (Barad 2012: 19). All actors of the exhibition performance are together, but they retain their integrity and individuality.

Laboratory flasks and devices transmattered into artworks are material records in which art, science, and technology are combined. As already observed, this transfer is often confusing to the audience, who tend to perceive liminal projects as too scientific.

Putting their collections on display, regular museums commonly place most exhibits in glass or plexiglass cabinets. Bioart sometimes uses both kinds of glass barriers, and this is mainly done to protect the items against the viewers' touch and even their breath. Living bacteria, yeast, plants, and cells are isolated to prevent them from becoming contaminated and from infecting the audience. At the same time, research materials are often put on show in display cabinets as unique records of the process, as exemplified by the exhibitions of my liminal projects.

The aestheticized protection against contamination can be fruitfully examined through the lens of Tsing's theoretical proposal, where contamination is framed as a basic form of collaboration. Consequently, the central question is whether the audience become contaminated despite the glass protection. This question is highly pertinent as contamination can be understood both in biological terms and as contamination by an idea, atmosphere, and emotion. The Killing Ritual conducted by SymbioticA is an exciting example of using contamination in its literal meaning as an integral element of the living show:

> At the end of every installation, we are faced with the ultimate challenge of an artist—we have to literally kill our creations. The works have to be terminated by the end of the show for both practical an conceptual reasons. For that we devised the Killing Ritual. The killing is done by taking the Semi-Living sculptures out of their containment and letting the audience touch (and be touched by) the sculptures. The fungi and bacteria which exist in the air and on our hands are much more potent than the cells. As a result the cells are contaminated and die (some instantly and some over time. (…) We also make a point to invite the people who invited us (curators, gallery directors, etc.) to participate in the killing, as they also are responsible for the existence of the semi-living sculptures presented in their show. On more than one occasion people from the audience have approached us after the ritual and admitted that initially they did not believe our sculptures were alive until they were killed. (Zurr 2008: 20)

As opposed to SymbioticA's deliberate action, biological contamination often occurs without artists' intention because flask and containers are not always perfectly sealed or because they are opened by mistake. It is also a common practice to dispose of Petri dishes with cells, bacteria, and other living agents after the exhibition closes. They have become contaminated by the 'outside' and as such cannot return to the laboratory. Killing a living exhibition is always ritualistic, even if it is not planned the way SymbioticA plans its Ritual Killings. This is done by the artists, curators, or other gallery workers. Petri dishes with liminal lives should not wind up in a regular rubbish bin because they are biohazardous material that should be neutralized by UV irradiation or hypochlorite. As an exhibition finishes, liminal lives lose their privileged position of living art producers, and they have to die and be disposed of exactly as it happens in the laboratory.

Living exhibitions are often accompanied by workshops. They offer an opportunity to incorporate the audience into the liminal projects on display. Workshops also play a narrative and educational role based on hands-on experiences they promote. Liminal practitioners share the laboratory methods used in their liminal projects with the participants. Almost all prominent exhibiting institutions have their education departments whose staff are responsible for developing show and workshop guidebooks. Liminal practitioners are sometimes involved in preparing the events, and sometimes they are not, but they are usually eager to run their workshops.

Workshops are, first of all, meetings and occasions to talk and to do something together with the audience. The idea of the project and its physical matter are transmattered by the collective activity.

Besides workshops associated with particular projects, there are also stand-alone workshops. In my liminal practice, I usually call them living experiments. *The Last Supper* performance was in a way a workshop that offered the participants an experience and food for thought. The scientific presentation that followed the consumption of beer and bread provided a platform for discussion. Workshops are as a rule performative, a quality that can later be presented in their photo and video documentation. They also give liminal practitioners an opportunity for collecting observations and obtaining immediate feedback, which rarely happens in exhibitions.

Arranging hands-on experiences comes with certain technical limitations. I could not conduct a cell-culture-based workshop for my *safe suicide* project. This would have required strict laboratory conditions, and random participants could not have been invited to the sterile work area. Additionally, tissue cultivating would have required basic training in simple laboratory procedures. Also, working in a laminar flow cabinet or pipetting media and reagents would have been far too expensive 'games' to offer to people to 'play.' Consequently, only conceptual workshops were possible in this project. The time factor made for another major difficulty as both my liminal projects took more than a year of work to develop. It would have been impossible to repeat the process even across multiple workshops. However, together with my main scientific collaborator, I conducted several laboratory life mediation and re-mediation workshops. To illustrate what they looked like, give the general idea of such liminal events, and look into their transmattering function, I present two scripts: one is mine and the other is by liminal practitioner Adam Zaretsky.

> *Constructing Life*
>
> The workshop is based on the dynamically developing field of technology—synthetic biology. It is a kind of molecular engineering that makes it possible to construct biological machines. It is based on the interdisciplinary cooperation of molecular biology, genetic engineering, bioinformatics, and systems biology. The primary goal of the workshop is to develop a narrative on synthetic biology and confront practice and scientific knowledge with its popularization in society.

> The first day will be dedicated to wet work and understanding synthetic biology methods. The group will work on transforming *E. coli*.
>
> The second day is going to be dedicated to narratives around synthetic biology and biotechnology in general. The group will be asked to read the book *Genes and Machines* before the workshop and to underline three kinds of passages in it: a) what is not understandable; b) what is infantilized; and c) what sounds controversial. The participants will also be asked to watch the documentary Genesis 2. 0.. Both assignments will be a basis for discussion.

The workshop took place in the same teaching laboratory that housed *The Last Supper* degustation. All the participants were affiliated with the humanities. Combining the analysis of the text and the film with simple biological experiments provided a good foundation for discussing synthetic biology and the general development of biotechnology. It was also important that the workshop took place in a scientific context, and the group had an opportunity to see what the laboratory looked like. Even though this activity was neither deeply emotional nor solidly exploratory, some participants were touched when they heard that we were to perforate bacteria's bodies. Attention then shifted to the ethical questions around the altering of bodies in and by science.

Zaretsky has carried out a broader educational and exploratory activity at the University of Leiden. As he announced:

> The Vivoarts: Biology and Art Studio course explores the intersection between art and biology and discusses the many cultural issues involved in bioart—and more generally in the field of life sciences—through a blend of hands-on laboratory protocols, critical reading, and the production of contemporary artwork. The ethics of producing living art are debated and made more tangible and understandable by the use of living material/organisms into the class final projects. (2009)

Zaretsky's intentions and goals were the same as in my workshop, but as the duration of his course was longer, more techniques were used, and more questions were explored. Though the venture was, on the face of it, purely educational, Zaretsky emphatically insisted that it differed from regular academic classes:

> The difference between a technical scientific learning session and a Vivo-artistic laboratory approach is mostly qualitative. While engaging in the

> technics, we also deal with the relational issues surrounding this type of process: pain, death, responsibility, curiosity, the meddlesome sadism of a personal genetic footprint/signature/graffiti, risk assessment between foreign species and the ecosphere, as well as critiquing admonition against the urge to fondle the folds of mutant love. (2009)

All the themes listed by Zaretsky as points of interest exceeded the scope of regular scientific research. First, there was an analysis of personal responses to and experiences concerning 'others.' There was also room for a critical scrutiny of scientific operations. Every participant had to transmatter their body by being involved in Zaretsky's 'living theatre.' He realized that scientists were suspicious about his classes: "I think the reaction to an art class in the lab doing 'important' transgenic embryology work in the name of non-utilitarian, 'frivolous' artistry reflects a fear of demystification of transgenic process" (2009). Such an awareness is vital for liminal practices, especially workshops and hands-on living projects. A participant in my *Constructing Life* workshop has written in her feedback form that she was surprised that all wet actions were so simple; she had expected them to be much more complex and undoable for non-scientists. Liminal activities transmatter science into something 'closer to life,' something to play with and to be engaged in. Zaretsky's course included some difficult moments connected to bioethical issues and simple human emotions. Students were encouraged to give names to the eggs they had fertilized, but then a moment came when they had to destroy their 'pupils.' It was essential to reflect on the daily laboratory activity of sacrificing, in its various forms ranging from autoclave and refrigeration to putting on ice or simply pouring down the drain (Zaretsky 2009). The workshop allowed students to perform this more respectfully and emotionally by administering a Valium overdose or through rituals of their own invention. That helped them to be really focused on the embryos as something that mattered.

Both hands-on experience opportunities depicted above were focused on experiencing and thinking. There were no objects that could be taken home or exhibited afterwards. These processual activities enabled participants to transmatter and be transmattered together with the workshop leader and all the 'invited' non-human beings.

## References


Barad, Karen. 2012. On Touching—The Inhuman That Therefore I Am. *differences: A Journal of Feminist Cultural Studies* 23 (3): 206–223. https://doi.org/10.1215/10407391-1892943. Accessed 31 July 2022.

Kember, Sarah, and Joanna Zylinska. 2015. *Life After New Media: Mediation as a Vital Process.* Cambridge, MA: MIT Press.

Kunst, Bojana. 2015. *Artist at Work, Proximity of Art and Capitalism*. Winchester, Washington: Zero Books.

Latour, Bruno. 1999. Pandora's Hope. In *Essays on the Reality of Science Studies.* Cambridge & Massachusetts & London & England: Harvard University Press.

Lowenhaupt Tsing, Anna. 2015. *The Mushroom at the End of the World: On the Possibility of Life in Capitalist Ruins.* Princeton, NJ: Princeton University Press.

Mackenzie, Louise. 2017. *Evolution of the Subject. Synthetic Biology in Fine Art Practice.* Newcastle: University of Northumbria.

Thacker, Eugene. 2004. *Biomedia.* Minnesota: University of Minnesota Press.

Zaretsky, Adam. 2009. Dangerous Liaisons and Other Stories of Transgenic Pheasant Embryology. *We Make Money Not Art.* https://we-make-money-not-art.com/_yes_its_true_im/. Accessed 31 July 2022.

Zurr, Jonat. 2008. *Growing Semi-Living Art.* The University of Western Australia, School of Architecture, Landscape and Visual Arts.

CHAPTER 9

# Conclusion: How Should One Transmatter?

A text has its limits—its beginning and its end—but this does not mean that material storytelling is contained. Designing and writing a text is somehow delayed vis-à-vis storied matter, which plots itself ceaselessly. The term I use—'trajectory'—is very vivid in terms of conveying the mechanism of selecting a fragment of material storytelling to be related and described. I selected three trajectories, determined by three culturally constituted contexts: the laboratory trajectory, the gallery trajectory, and the academic trajectory. These contexts worked as theatres where material stories were performed/enacted before they were written down. Without the contexts, making a choice of what should be described would be very tricky.

While all researchers engage their bodies in the inquiry process, they differ in terms of the forms of their embodied engagement and in the ways their participation is present in their written stories. Some people use their bodies as tools for sitting up to read, write, and re-write for hours on end. Fieldwork researchers travel around and collect observations and samples. Finally, experimental scientists use their perception and senses to examine the material world. The common point for all these individuals is telling a story about processes, relations, and intra-actions.

This is the moment when I must return to the fundamental question I asked at the beginning of this text: How should one transmatter? The question requires drawing up a recipe—or at least a list of guidelines—in



K. Żyniewicz, *The Body as Matter in Art and Science*,
https://doi.org/10.1007/978-3-032-13665-7_9

this chapter. Some additional questions cropped up when this text was being written and read (by myself and by other readers): Why are the operations done on matter and described in the text called transmattering, instead of transformations? Is transmattering individual or collective? Can matter be not transmattered? Who can do transmattering? I can't predict every possible question that can be asked by a reader. Even less achievable is answering all the possible questions. As the notion of transmattering is the basis of this text, its understanding is crucial. The notion must be strong and flexible simultaneously. First and foremost, it must have solid underpinnings. When translated into Polish, transmattering means a transformation of matter. There is no terminological conflict between transformation and transmattering. Transmattering is more than transformation, and its definition is based on how transformation is understood in biology. To transmatter means to transform on several different levels at the same time. In my research, the co-vibration with three other areas and contexts enabled me to transmatter physical matter and its meaning through those three contexts simultaneously.

For historical reasons, the word 'transformation' has two meanings in biology. In genetic engineering, transformation means introducing DNA into the cells of microorganisms (Green and Sambrook 2012). Such a transformation can be a natural and spontaneous process (as happens in certain bacteria), or it can be induced by a variety of methods (in *The Last Supper* project, I used chemical treatment of yeast cells when I transformed them with a plasmid containing my gene). The analogous process of introducing DNA into plant and animal cells is called transfection because the term 'transformation' is used to refer to a malignant transformation, where normal cells turn into cancer cells as a result of genetic changes caused by internal or external factors (Hanahan and Weinberg 2000). Despite these differences, both meanings, of transformation denote change in the genetic information of a living cell. In biological transformation, living matter changes some of its properties but remains the same cellular entity on the fundamental level. Moving between the living body and laboratory liminal beings, such as isolated cells and genes, between different meanings and epistemic cultures, and spanning the entire continuum of life and death, my work has been much more than this biological transformation and called for a new term—transmattering. I did conduct a malignant transformation when immortalizing my cells, but it was one of the multiple operations on matter that were performed in my research. The terminology of the laboratory transmattered the general cultural discourse on life, death, and immortality.

Cancerous cells associated with a deadly illness in medicine have already been portrayed and presented as artistic objects. Transfection also occurred in my trajectories, as I introduced my gene to yeast cells. This was not the end of the process, though, and the transgenic yeast produced alcohol, which was consumed by the participants in my artistic performance. On the metaphorical level, my body was transmattered into a food product. I am listing all these operations to picture the complexity of transmattering. Interpreted from the perspectives of art and academia, biological transformations acquired new and unexpected meanings. The same happened to biology when scientific experiments were subjected to external interpretations. For its part, the complexity of transmattering meaningfully influenced the structure of this text. It may look complex, even chaotic at times, but it is difficult to depict the intensity of liminal co-vibration in a text.

The question about the individual and collective dimensions of transmattering is essential. I believe that readers can find it challenging to determine whether the trajectories are described as individual stories or as collective experiences. Using autoethnography suggests that the research was carried out and described by me, an individual of unique embodiment and equally unique situatedness in the world. However, I have never been alone, and it has always been I-with. In all my activities, I have been in entanglements. My existence as such is founded on being-with and being in-between. As a liminal being, I have co-vibrated with networks of agents in three different zones, putting them somehow in touch with one another. Even writing this text, though based on my decisions and formulations, was not only mine—all the networks with which I co-vibrated were writing and reading with me.

Consequently, this text is not addressed to one group of readers, and it can be read by an array of people from art, biology, and the humanities. I can also imagine that every reader, being-with others, will read-with them what I wrote about the personal matter of I-with. It is not only biological matter and not only my body, and it comprises others—humans, non-humans, and post-humans that have existed in co-vibration with me. The personal matter is the matter of an entangled person.

No matter exists in stagnation. All matter is permanently changing and constantly transforming. However, not every matter must be transmattered. Transmattering requires particular circumstances and complexities.

Not every matter is transformed by several contexts at the same time. Not every body is transformed by isolating its parts and changing their status. It is rather unusual for bodies to be simultaneously used for laboratory experiments and art. In terms of transformation as defined in biology, matter is transformed through biological matter or re-mattered through knowledge. In this chapter, I seek to provide a protocol similar to those I used in my laboratory trajectory. Like in laboratory work, the protocol does not guarantee obtaining the expected result. It just spells out the way and helps to structure it. Who can do transmattering? I would say that all those who are eager to become liminal and experiment with them-with, biologically, mentally, socially, and artistically. First of all, not everybody must transmatter. It is not the only cognitive path available to humans.

Since my research was based on three fields that transmattered each other, I thought that a protocol similar to those used in the life sciences would work best. Biologists follow protocols or draw up new ones after successful experimentation. Protocols explain what should be done to repeat an experiment accurately and see if it works. The first important thing is to be sensitive and open to exploring the different forms and contexts of matter. Every I-with must be aware of its complex material structure and learn how to use it in transmatterive research. The researcher's body is a crucial component of I-with, responsible for selecting stories from storied matter. However, in order to discern a material story, the body should be responsible and response-able. I-with possesses enough competence to write a personal story but simply being I-with is not enough to compose an academic story that meets requirements for research writing. The academic context makes I-with consider its own experiences and other stories produced in this context. This can be viewed as an opportunity or as an obstacle. The advantage of knowing the already told stories is that one can enter into a dialogue with them, instead of simply repeating them. The necessity to know theoretical matter can also be regarded as a limitation because it forces the voice of I-with to confront other voices time and again. As a result, a researcher may feel that they are never credible enough without quoting other stories. Transmattering happens everywhere, so academia is not the only site where it can be carried out and described. Every I-with can decide about the context of plotting a transmatterive story.

Storied matter is replete with themes. It is challenging to find the right topic and reason for conducting transmatterive research. It is a matter of asking pertinent questions. The initial questions can be very abstract, and they should stimulate the imagination and initiate creativity. The next step involves the selection of physical matter to be explored, observed, and analysed with a view to answering the initial questions. Practical experiences are crucial in transmatterive research. Physical transformations breed theoretical reflections and transform meanings and their expression in language.

Given that both liminal projects explored in this text were based on the matter coming from my body, one sound question to ask is whether using pieces of one's own body is a prerequisite of transmatterive research. The answer is that it is not. Since the human body is entangled in I-with, it is not the only available matter to be transmattered. Whether working with bacteria, plants, or anything else, the human body is anyway engaged in research. What is far more important than using (or not using) one's own body as a research material is experimental working on physical matter. If the laboratory is not the only possible place where this can occur, it is an exciting context to consider. The so-called 'wet lab' helps us to better understand the construction of I-with, which is difficult to grasp based on theoretical knowledge alone. The combination of experimental fieldwork and theoretical inquiry is my recommended recipe for transmatterive research.

Adopting a liminal position and exploring several contexts simultaneously are among the essential points of my protocol. In my research, I tested co-vibrating with three fields of cultural production, and this was challenging, especially when it came to writing about it. How many fields can be approached simultaneously is difficult to determine. I believe that the more significant the network is, the more challenging co-vibrating becomes. Liminality gives one an opportunity to be flexible and open to many forms of matter and stories based on it. Cognitive co-vibrating with many fields of cultural production encourages bypassing rigorous schemes applied in strictly defined areas. Under such circumstances, one is likely to put together theoretical proposals that sound blasphemous to particular disciplines. Rule-breaking lies at the core of liminal transmatterive research. Such research requires accepting imperfection (even more than disciplinary research does). A liminal inquiry and a liminal text are both patchwork-like. When engaged in transdisciplinary co-vibration with I-with, people with a fixed disciplinary affiliation are bound to be unsatisfied because liminal research will use their fields fragmentarily, without fully considering their traditions and knowledges. Not everybody will like this mode of doing research and writing a text. Also, not everybody will know how to understand liminal research. Like all forms of knowledge production, this framework is only one of a range of available options. I by no means suggest that liminal research is the only feasible way to study the contemporary world.

Writing in the first person was vital for my transmattering operations. I believe that this approach is indeed crucial if one wants to conduct transmattering. Even if a researcher has not worked with physical matter taken from their body, writing on behalf of I-with gives the impression of actual embodied involvement in all the processes. In this way, it is made clear that the researcher is a storyteller who composes a story of material transformations. Autoethnography is an academically defined and recognized method. I adopted it both in my academic research and writing because, even if governed by its specific rules, the method is also remarkably flexible. However, not all personal experiences must be called autoethnography. Personal experiences as such can be incorporated into a text/story to show or explain particular phenomena. I conducted autoethnographic observations for more than four years and kept fieldwork journals to record them, but citing one event in one's research or life can work just as well for other people. It is a matter of individual decision.

The story I have plotted brings together three contexts—laboratory, art gallery, and academia—as characteristic of my autoethnography. I believe that the contextual network can be different for other practitioners. The most important thing is not to remain in one context simply because that would make the prefix trans- unwarranted. The inquiry process is based on establishing/constructing an order for many ideas and experiences. It is crucial to co-vibrate with the selected field with full engagement and openness. Research is usually very intense and extensive, and this is its most significant value. It makes sense to be brave and fully immersed in the simultaneous co-vibrating with all the fields involved, without being preoccupied with the limitations of systems and structures. Meeting these limitations, facing education systems is painful, but it should not be seen as an insurmountable obstacle. Limits are part of all inquiry. Without them, researchers would easily become lost in research processes. The system requires constructing a framing for the story, which makes one reflect and answer critical questions, such as: What did I experience? What does it mean? How am I going to share it with the world? How does my experience relate to experiences described by others? These questions help one to look back and then to look forward. Selecting particular trajectories is like stopping to take a breath while running a marathon. I-with is not only a collaborative existence from a biological perspective. It is also 'with' in a social sense. This means that all research is entangled in the network of other instances of research. Freedom is limited by being-with. Acknowledging that is very helpful.

Another thing that should be acknowledged and accepted is that liminal research is homeless. It is always temporarily located in particular fields, mainly because this is what the structural divisions of the academic research system make mandatory.

I have presented my research at transdisciplinary meetings and conferences dedicated to contemporary art, bioethics, and even biology. Liminal research is always partly correlated with each field that it touches. The word 'partly' is critical. Complete accommodation within any given field would mean losing the liminal positioning. A liminal researcher must be ready to face obstacles resulting from being a stranger in all structured fields. My liminal research is too artistic for academia and too scientific for the artworld. When co-vibrating only takes place between different academic fields, the situation in which practitioners find themselves is probably somewhat easier. However, liminal researchers should be prepared to constantly explain why their research is liminal and why it is partially adjusted to several cultural zones.

I decided to share my autoethnographic experiences from biological laboratories in the field of art and in academia. In each of these cases, I needed to use a different language. Co-vibrating with academic fields resulted in textual production, while displaying liminal laboratory projects at art galleries required inventing a visual and metaphorical language. It was a matter of my methodological decisions. However, it is possible to carry out transmatterive operations and share insights into them either in academic research or in art alone. Methodological choices determine how long the examined trajectories will be and how many transmatterive moments will be considered. Since co-vibrating with several fields can be overwhelming and chaotic, I strongly recommend structuring the research design with a lot of precision. Having a clear research structure in mind will help practitioners to avoid becoming lost in the massive maze of knowledge.

As emphasized before, the trajectories I selected for analysis in this text are only links in the continuity of my practice. They are grounded in previous transmatterive trajectories, and they serve as a basis for future trajectories. This continuity is essential. Even if a researcher works on a research topic for some time and then shifts to something very different, I-with is already transmattered. Being conscious of that and learning how to use prior trajectories in experiencing new ones are pivotal to any liminal project. The protocol offered above is less meticulous or detailed than regular laboratory protocols. I could not give either particulars of or obvious instructions on what should be done step by step in transmatterive research. Instead, I listed all the points I believed crucial to structuring a trajectory similar to mine.

My answer to the fundamental question—How should one transmatter?—is incomplete and somewhat tentative, and I accept this imperfection because it can inspire other I-withs to be imperfect. Accepting imperfection is crucial in experimentation, as I learned from the biologists at the laboratory where I developed *safe suicide* and *The Last Supper*. This text is matter to be transmattered by others. All its defective elements are gaps to be filled by their transmatterive investigations.

## References

Green, Michael R., and Joseph Sambrook. 2012. *Molecular Cloning: A Laboratory Manual*. New York: Cold Spring Harbor Laboratory Press.

Hanahan, Douglas, and Robert A. Weinberg. 2000. The Hallmarks of Cancer. *Cell* 100 (1): 57–70. https://doi.org/10.1016/S0092-8674(00)81683-9. Accessed 1 August 2022.

# Appendix 1: Laboratory Notebook: *Safe Suicide*

**15 March 2016**—separation of B lymphocytes from fresh blood/immortalization by Epstein-Barr virus (protocol by Elona)

1. Add 10 ml of MSL solution to a 50 ml tube.
2. Dilute blood 1:2 with PBS or RPMI 1640.
3. Add the diluted blood very carefully on top of MSL solution.
4. Spin 30 min/1000 g.
5. Collect the plasma in clean tubes.
6. Collect a layer of MNC (mononuclear cells) into a clean tube.
7. Wash with 10 ml of PBS or RPMI 1640.
8. Centrifuge 10 min/400 g.
9. Discard supernatant.
10. Wash again with 10 ml PBS or RPMI 1640.
11. Centrifuge 10 min/400 g.
12. Discard supernatant.
13. Work on ice:

    Resuspend the pellet in 1 ml of freezing solution (take 10 μl of solution and count lymphocytes using a Burker cell counting chamber).

    Freeze the solution at −20 °C for min. 12 h, at −80 °C for min. 12 h, and then transfer to liquid $N_2$.

    Working with blood must happen on the same day.



K. Żyniewicz, *The Body as Matter in Art and Science*,
https://doi.org/10.1007/978-3-032-13665-7

1. Transfer the blood (5 ml) to test tubes with sodium citrate.
2. Mix the blood gently by shaking upside down.
3. Transfer the blood to LeucosepR 163288 test tubes (Greiner Bio-one) with separating medium and centrifuge 800 g/15 min at room temperature. You need to obtain six layers: plasma, lymphocytes and mononuclear cells, separating medium, a porous layer, the separating medium, and erythrocytes and granulocytes sediment.
4. Remove the plasma layer.
5. Take the lymphocytes layer gently using a Pasteur pipet and transferred it to a new test tube (Falcon).
6. Add 10 ml of PBS (Sigma-Aldrich) and centrifuge 10 min/250 g. Repeated twice.
7. Make a 1 mg/ml dilution of Cyclosporine A in alcohol.
8. Dilute the cyclosporine A dilution in RPMI-1640 medium to obtain the 1:50 proportion (cyclosporine A: RPMI-1640).
9. Dilute the cellular pellet in 4 ml of the completed RPMI medium.
10. Add 1 ml of EBV (Epstein Barr Virus) pellet and 50 µl of the Cyclosporine A dilution.
11. When ready, transfer the solution to a cultivating bottle with filter and put into an incubator at 37 °C and 5% $CO_2$.

¼ of cells were left without virus, and ¾ of cells were infected with the virus and left for about one week.

**23 March 2016**—3 ml of RPMI medium added to the bottle with the non-immortalized cell without immortalizing has been added. The immortalized cells left without changes. Clumps started to be visible.

**25 March 2016**—counting the cells using a Thoma chamber, staining with trypan blue.

**30 March 2016**—3 ml of RPMI medium added to the bottle with non-immortalized cells.

3 ml of RPMI medium added to the bottle with the immortalized cells for 0.5 h; then discarded 3 ml of the medium. In the end, Cyclosporine A was added (to help the virus to enter the cells).

**31 March 2016**—counting the non-immortalized cells using a Thoma chamber: 906250/ml.

**4 April 2016**—counting the non-immortalized cells using a Thoma chamber: 1812500/ml.

**6 April 2016**—changing the medium in the immortalized cells: 3 ml of medium removed,

3 ml of new medium added; clumps are visible.

**7 April 2016**—counting the non-immortalized cells: 125000/ml; possible mistake: the immortalized cells counted instead of the non-immortalized cells.

**11 April 2016**—counting the non-immortalized cells: 531250/ml.

**14 April 2016**—counting the non-immortalized cells: 1468750/ml.

**19 April 2016**—passage of the immortalized cells, the last change of the medium in the non-immortalized cells.

**22 April 2016**—passage of the immortalized cells.

**25 April 2016**—passage of the immortalized cells (8 bottles).

**28 April 2016**—freezing the cell line:

1. Centrifuge for 5 min.
2. Discard supernatant.
3. Prepare a freezing solution: 100 µl of DMSO + 200 µl of FBS.
4. Add to 700 µl of medium to the cells and move them to test tubes with the freezing solution.
5. Put the test tubes into a freezing box and move it to a freezer.

**14 May 2016**—preparing immuno-specimens:

1. Fix the cells in 4% PFA solution at room temperature for 10 min.
2. Wash the cells 3× in PBS solution.
3. Incubate the cells in 0.1% Triton × PBS solution at room temperature for 3 min, in order to permeabilize the cell membranes.
4. Incubate the place of non-specific tie, 0.25% glycine (Sigma Aldrich) and PBS solution at room temperature for 30 min.
5. Incubate the place of non-specific tie in BSA (Sigma Aldrich) + PBS + 0.05% Tween (Sigma Aldrich) solution at room temperature for 1 h.
6. Incubation with I—row antibody at 4 °C overnight.
7. Wash the cells 3× with PBS.
8. Incubation with II—row antibody at room temperature in the dark for 90 min.
9. Wash the cells 3× with PBS.
10. Stain the nuclei with Draq 5 (Biostatus Limited), 5 min incubation, Draq 5 resolved in PBS 1:1000, room temperature.

**17 May 2016**—defrosting the cells:

1. Prepare a falcon tube (15 ml) with 5 ml of medium.
2. Move the cells to the falcon.
3. Centrifuge the cells for 5 min.
4. Discard supernatant.
5. Add 5 ml of fresh medium and centrifuge the cells again.
6. Prepare cultivating bottles with 8 ml of medium in each.
7. Discard supernatant.
8. Resuspend the pellet in 2 ml of medium and add 1 ml to each of the bottles.

**19 May 2016**—passage of 4 bottles:

1. Move the cells to falcons (15 ml).
2. Centrifuge the cells (5 min).
3. Discard supernatant.
4. Add 8 ml of medium to big bottles and 5 ml to small bottles.
5. Resuspend the pellet in 2 ml of medium and add 1 ml to each of the bottles.

**22 May 2016**—counting the cells: 687500/ml.
**24 May 2016**—RNA isolation:

1. Centrifuge the cells (10 min).
2. Discard supernatant.
3. Add 800 µl of TRIzol.
4. Resuspend the pellet and move 400 µl to 2 test tubes.
5. Add 80 µl of chloroform to each of the test tubes and shake strongly.
6. Add 200 µl of the solution of Phenol, chloroform, and iso-amyl alcohol.
7. Centrifuge (10 min).
8. Add 200 µl of the solution of Phenol, chloroform and iso-amyl alcohol.
9. Centrifuge (5 min).
10. Precipitation (RNA separation).
11. Add 500 µl of 96% Ethanol and 70 µl of sodium acetate.

**25 May 2016**—freezing the cells; one bottle left in cultivating, having blood taken, and the isolation of B lymphocytes to leave them without immortalization.

**30 May 2016**—changing the medium in the cells, preparation of polylysine slides:

1. Wash slides with 70% Ethanol and sterile PBS, put them in the vertical position to dry in a laminar flow cabinet, and sterilize them with UV (15 min).
2. Add polylysine solution to a sterile Petri dish; the slides should be submerged in the medium on both sides.
3. Put the slides into the medium; make sure that the slides do not stick to each other or to the bottom of the petri dish.
4. Incubate the slides in the medium in an incubator for 60 min for cultivating (37 °C).
5. Remove the slides from the liquid and let them dry in the vertical position.
6. Store the slides in a sterile petri dish, at room temperature.
7. Store the polylysine solution in a sterile falcon at −20 °C.

Polylysine solution: 25 mg of Poly-L-Lysine resolved in 50 ml of dd$H_2O$.

**31 May 2016**—counting the cells (one week after isolation): 156250/ml.

**2 June 2016**—counting the cells: 125000/ml.

**3 June 2016**—freezing the cells from the big bottle—3 test tubes, 1 ml taken to experiment with bacteria, 6 ml left in cultivating.

**6 June 2016**—counting the cells: 31250/ml; preparation of the cells for an experiment with *Listeria monocytogenes* and *Staphylococcus.*

**7 June 2016**—the cells cultivated in a multi-wells petri dish through the night did not survive.

**8 June 2016**—freezing the immortalized cells; counting the non-immortalized cells: 31250/ml.

**13 June 2016**—counting the cells (non-immortal): 31250/ml.

**15 June 2016**—preparing the next bottle of medium: 5 ml of HEPES, 5 ml of penicillin/streptomycin, and 50 ml of FBS; cultivation on polylysine slides, 300 μl of medium with cells for each well.

**18 June 2016**—fixing the cells on polylysine slides:

1. Add 3% PFA for to the wells for 10 min.
2. Wash the slides 3× with PBS; leave the slides in the last liquid.

**21 June 2016**—defrosting HeLa cells:

1. Warm the cells up by holding the test tube in your palm.
2. Move the cells to a falcon (15 ml) and add 5 ml of DMEM medium + 10% FBS.
3. Centrifuge it (3 min).
4. Discard supernatant.
5. Resuspend the pellet in 5 ml of medium and centrifuge it again (3 min).
6. Prepare a 10 ml petri dish.
7. Discard supernatant, resuspend the pellet in 1 ml of medium, and move it to the petri dish.

The joint cultivation of my B lymphocytes and HeLa cells:

A petri dish with 10 ml of RPMI medium, 1 ml of lymphocytes, and 1 ml of HeLa cells

**23 June 2016**—counting the cells: 28125/ml.

**27 June 2016**—HeLa—without changing the medium, lymphocytes—changing the medium, HeLa + lymphocytes—changing the medium.

**29 June 2016**—peeling HeLa cells:

1. Add 2 ml of Trypsin and wait about 10 min.
2. Add 2 ml of medium and move the cells to a falcon.
3. Centrifuge it (3 min).
4. Add a fresh portion of medium to the petri dish.
5. Resuspend the pellet in 1 ml of medium and move it to the petri dish.

HeLa—changing the medium

Lymphocytes—changing the medium

HeLa + lymphocytes—changing the medium without peeling HeLa

**30 June 2016**—an experiment with Listeria monocytogenes and Staphylococcus:

1. Add 1 ml of medium with cells to 10 wells and leave it for the night.
2. Prepare a bacteria dilution: 5 test tubes—900 μl of PBS.
3. Add 100 μl of the bacteria to the first test tube and then transfer 100 μl from one test tube to another.
4. Prepare another 2 test tubes with 900 μl as a control.
5. Add 100 μl of the bacteria to 10 wells and incubate for 1 h.

**30 June 2016**—formaldehyde-fixed cells:

1. Wash the cells twice with prewarmed phosphate-buffered saline, pH 7.4 (PBS).
2. Fix the sample in 3.7% formaldehyde solution in PBS at room temperature for 10 min. Note: Methanol can disrupt action during the fixation process. Seeing this, it is best to avoid any methanol-containing fixatives. Methanol-free formaldehyde is the preferred fixative.
3. Wash 2× or more with PBS.
4. Place each coverslip in a glass petri dish and extract it with a solution of acetone at 20 °C or with 0.1% Triton X-100 in PBS for 3 to 5 min.
5. Wash 2× or more with PBS.
6. When staining with any of fluorescent phallotoxins, dilute 5 μl of the methanolic stock solution into 200 μl PBS for each coverslip to be stained. To reduce nonspecific background staining with the conjugates, add 1% bovine serum albumin (BSA) to the staining solution. It can also be useful to pre-incubate the fixed cells with PBS containing 1% BSA or with the Image-iTTM FX signal enhancer (I36933) for 20–30 min prior to adding the phallotoxin staining solution. When staining with biotin-XX phalloidin (B7474), dilute 10 μl of the methanolic stock solution into 200 μl PBS for each coverslip to be stained. When staining than one coverslip, adjust the volumes accordingly. For a stronger signal, use 2 or 3 units per coverslip.
7. Place the staining solution on the coverslip at room temperature for 20 min. To avoid evaporation, keep the coverslip inside a covered container during the incubation.
8. Wash 2× or more with PBS.
9. When using biotin-XX phalloidin, incubate for 30 min with 100 μl of the 10 μl/ml solution of fluorescent or enzyme-conjugated

streptavidin dissolved in 100 mM Tris-HCL, pH 7.5, 150 mM NaCl, 0.3% Triton X-100, and 1% BSA. Incubate at room temperature for 15 min. After incubation, wash the coverslip with PBS. To stimulate enzyme activity, follow the procedure recommended for a given enzyme.

10. For long-term storage, the cells should be air-dried and then mounted in a permanent mountant, such as Pro Long Gold reagent or Cytoseal. Specimens prepared in this manner retain actin staining for at least six months when stored in the dark at 2–6 °C.

**31 June 2016**—passage of HeLa cells to two petri dishes (one with RPMI medium and the other with DMEM medium).

**11 June 2016**—observation of joint cultivation (my immortalized B lymphocytes + HeLa cells).

**15 June 2016**—cultivating a combination of cells on polylysine slides; freezing a portion of the cells; preparing a portion of the cells for a rotenone experiment.

**19 July 2016**—lymphocytes + HeLa on RPMI—5 µl of rotenone
lymphocytes + HeLa on DMEM medium—2.5 µl of rotenone
lymphocytes + HeLa on a mix of media—control

**20 July 2016**—Lymphocytes + HeLa on a mix of media—2.5 µl of hypochlorite

**11 August 2016**—defrosting one test tube for RNA isolation; cultivating the cells in 2 bottles (with 6 ml of medium in each).

**16 August 2016**—counting the cells before RNA isolation: 5625/ml; repeating the experiment with bacteria:

1. Prepare the solution: add 1.3 ml of 4 ml of stock to 18.7 ml of medium and add it to 6 wells (3 ml of medium per each).
2. Preparing the bacteria:
   - prepare flasks with 3 ml of medium in each;
   - heat the asa with fire;
   - take a colony of bacteria from the petri dish and add it to the fluid medium in flasks; mix gently;
   - heat the flask inlet and asa;
   - close the flasks carefully.

**17 August 2016**—carrying on the work with bacteria:

1. Use the spectrometer to measure the density of the bacteria.
2. Move the solution with the bacteria to test tubes (1 ml per each).
3. Discard supernatant; add 1 ml of PBS.
4. Wash the wells twice with PBS.
5. Leave the multi-wells petri dish in an incubator for 3 h.
6. Discard the medium; wash the slides 2× in PBS and once in PFA.

**9 September 2016**—contamination of the cultivation with yeast.

**13 September 2016**—the beginning/onset of lymphocyte cultivation.

2 bottles—10 ml of RPMI medium, application of interleukin 1

**16 September 2016**—counting the cells using a Neubauer chamber: 12275/ml.

**19 September 2016**—counting the cells using a Neubauer chamber: 16600/ml.

**20 September 2016**—preparing the cells for an experiment with doxorubicin and etoposide administration:

1. doxorubicin: dilution 1:50 (1 µl-doxorubicin: 49 µl -PBS)
   - 1 well—9 µl
   - 1 well—18 µl
2. etoposide: dilution 1:100 (1 µl-etoposide: 99 µl-PBS)
   - 1 well—7.5 µl
   - 1 well—22.5 µl (7.5 × 3)
3. Fixing the specimens:
   - Sow the cells on bigger slides in a 6-wells petri dish, 400000 per well.
   - Wash 2× with PBS (1 ml).
   - Fixing with PFA—at room temperature for 10 min.
   - Wash 2× with PBS.
   - Closing the specimens with Dako/permount medium.
   - Storage in a fridge overnight.
   - Using of nail polish to cover the specimens.

Preparing PFA: warm 225 ml of water in a microwave; add 50 µl of 10N NaOH (40 g/100 ml) and 10 g of PFA in sterile conditions; mix gently to obtain a transparent solution; add 250 µl of 2 M HCl and 25 ml (10xPBS); store the solution at −20 °C; after defrosting, keep the solution at 4 °C, not longer than 2 weeks.

**21 September 2016**—repetition of the experiment with doxorubicin and etoposide:

1. doxorubicin 20 μl M: 1:10 60 μl/ ml PBS
2. etoposide 20 μl M: 1:100 4 μl/199 μl PBS

**10 October 2016**—preparing of cells for a UV experiment; defrosting:

1. Move the cells from the freezing test tube to a falcon (15 ml) with 5 ml of medium.
2. Centrifuge for 10 min.
3. Discard supernatant.
4. Add 2 ml of medium and resuspend the pellet.
5. Prepare 7 ml of medium in the bottle and add 1 ml of dilution from the falcon.

**12 October 2016**—passage of the cells.
**14 October 2016**—passage of the cells.
**17 October 2016**—counting the cells: 547500/ml.
**18 October 2016**—the UV experiment:

I trial—up to 10 Jls/$cm^2$
II trial—up to 15 Jls/$cm^2$
6 ml PBS + 1.5 ml of medium with cells

**19 October 2016**—preparing specimens using funnels:

1. Put 2 ml of medium with cells on basic slides through a funnel.
2. Shake for 5 min.
3. Remove the funnel and outline the lasting drop with the formaldehyde marker, leaving some margin around.
4. Soak the drop in using a paper towel.
5. Fill the funnel with PBS (not necessarily sterile).
6. Use parafilm to cover the funnel and keep it in freezer overnight.
7. After 24 h, take the slides out of PBS.
8. Make a dilution of Dapi (1:3).
9. Drop Dapi on the slides and keep in the dark for 10 min.
10. Put the slides back to the funnel with PBS for 3 min.
11. Replace PBS with alcohol and incubate it for 3 min.

12. Add a drop of closing liquid on the basic slides and put the cover slides on.
13. Remove the excess ofliquid and use nail polish to close the specimen.
14. Leave the specimens to try.

**24 October 2016**—biopsy of my skin and the beginning of fibroblast cultivation:

1. Pour PBS with antibiotic and bioptat on a petri dish.
2. Separate the epidermis from the main skin.
3. Move a piece of the skin to a new petri dish and leave it to dry.
4. Add the medium drop by drop, really gently to keep the skin piece stuck to the bottom of the petri dish.

**28 October 2016**—checking if the skin is still stuck to the bottom; moving the floating bits to a fresh petri dishes.

**2 November 2016**—observing the migration of the fibroblasts from bioptat; changing the medium

**4 November 2016**—the first passage of the fibroblasts:

1. Discard the medium.
2. Wash the petri dish with PBS.
3. Add 3 ml of Trypsin and put the petri dish into an incubator for 10 min.
4. Add the medium, especially to the place with the explant.
5. Centrifuge for 5 min.
6. Discard half of supernatant and replace it with a new portion of the medium.
7. Move the solution to a petri dish.

**7 November 2016**—changing the medium, observing the fibroblast migration.

**10 November 2016**—changing the medium, observing the fibroblast migration (visible in petri dishes 1 and 2).

**14 November 2016**—explant removal.

**16 November 2016**—changing the medium.

**17 November 2016**—petri dish 1: cell passage; petri dish 2: changing the medium.

**21 November 2016**—petri dish 1: changing the medium; petri dish 2: removed.

**23 November 2016**—changing the medium.

**25 November 2016**—passage in petri dish 1 and changing the medium in petri dish 2.

**2 December 2016**—passages: 2 and 3.

**5 December 2016**—passages: 4 and 5.

**7 December 2016**—changing the medium.

**9 December 2016**—freezing all the passages (2, 3, 4, and 5).

**12 December 2016**—defrosting passage 2.

**14 December 2016**—passages 2 and 4 for freezing; passage 5 for passaging.

**16 December 2016**—passaging passages 3 and 6 (two bottles).

**19 December 2016**—preparation of staining (BrdU, B-gal, 53BP1):

Passage 3:

1. Resuspend the cell pellet in 3 ml of medium.
2. Add 1 ml of medium.
3. Use a 12-wells petri dish and add 1 ml of medium to each of the wells (for BrdU and 53BP1).
4. Use a 24-wells petri dish and add 0.5 ml of medium to 2 wells (B-gal).

Passage 6:

1. Resuspend the pellet in 10 ml of medium.
2. Use a 12-wells petri dish and add 1 ml of medium to 9 wells (BrdU, 53BP1).
3. Use a 12-wells petri dish and add 0.5 ml of medium to 2 wells (B-gal).

**20 December 2016**—preparation of staining:

BrdU: 6 μl of BrdU per well (1.5 ml of medium)

**21 December 2016**—preparation of staining:

B-gal:

1. Discard PBS and replace it with fresh PBS.
2. Pour the closing medium, 1 ml per well (54 μl of formaldehyde, 25 μl of glutaraldehyde 8%, and 921 μl of PBS0; vortex the solution gently).
3. Incubate at room temperature for 5 min.

4. Wash 3× with PBS.
5. Pour the staining liquid, 1 ml per well (650 µl of water, 200 µl of buf.Fosf pH=6, 25 µl of cyanide 1, 25 µl of cyanide 2, 30 µl of NaCl, 2 µl of $MgCl_2$, and 50 µl of x-gal; vortex the solution gently).
6. Cover the petri dish with aluminium foil and leave in an incubator for 18–24 h.
7. Wash wells after incubation 2× with PBS; leave the last pouring.

DAPI staining: 1:3 (0.5 ml DAPI:1 ml PBS)
BrdU: washing 3× with PBS, pouring with alcohol, storage at −20 °C
**2 January 2017**—Immuno BrdU:

1. Add BrdU (from stock 4 µl/ml) to medium.
2. Fix the cells with alcohol (70% EtOH) and store them at −20 °C for about 1 h.
3. Wash the cells with PBS (5 min).
4. Wash the cells with PBS+0.5% Triton, 2 × 5 min (50 ml PBS + 2.5 ml Triton).
5. Pour the specimen with 0.5 ml of PBS and add 0.5 ml of 4 N HCl. Incubate for 30 min.
6. Wash with PBS 2 × 5 min.
7. Incubate with 1 ml of Borex 0.1 M per 1 min.
8. Wash with PBS 2 × 5 min.
9. Incubate with I-row antibody, anti-BrdU 10 µl/ml in PBS + 0.5% Tween for 1 h.
10. Wash with 2× PBS + 0.5% Tween (50 ml PBS + 2.5 ml Tween).
11. Incubate with II-row antibody (anti-mouse) 1:500 in PBS + 1%BSA + 0.5% Tween for 1 h.
12. Wash 2 × 5 min with PBS + 0.5% Tween.
13. Incubate with DAPI in the dark for 15 min (1:3.9 ml PBS + 3 ml DAPI).
14. Wash 3 × 5 min with PBS.
15. Close the specimens.
16. Observe the specimens after 24 h.

53BP1—DNA damage:

1. Fix the cells with 2%PFA in RT for 20 min.
2. Wash the cells with 10 ml of PBS; centrifuge for 5 min.

3. Discard supernatant, wash with 5 ml of PBS, and vortex.
4. Discard supernatant, resuspend the pellet in 0.5–1 ml of PBS, and vortex.
5. Add 120 μl on slides.
6. Cytospin, 800 rpm/80 g for 4 min.
7. Move the specimens to Coplin's jar and pour PBS; wash for 15 min.
8. Move the specimens to Coplin's jar which has been stored at −4 °C.
9. Add cold 70% ethanol in RT for 20 min.
10. Move the specimens to a new jar; wash with PBS for 15 min.
11. Block in RT for 30 min (wet chamber).

**4 January 2017**—microscope observations.
**6 January 2017**—passage 5.
**9 January 2017**—passage 6.
**13 January 2017**—defrosting passage 4.
**17 January 2017**—the phalloidin experiment:

1. Fix the cells with 2% PFA, wash 3× with PBS, and leave the cells in the third washing.
2. Wash the cells with PBS.
3. Permeabilize with 0.1% Triton X-100 in PBS for 30 min (50 ml PBS + 0.5 ml Triton).
4. Wash 2–3× with PBS.
5. Incubate for with phalloidin (1:40) for 3 min.
6. Wash 2–3× with PBS.
7. Close the specimens with DAPI.

**20 January 2017**—changing the medium in passage 4.
**23 January 2017**—passage 5—counting: 117187.5/ml.
**25 January 2017**—changing the medium.
**27 January 2017**—passage 6—counting: 476563/ml.
**30 January 2017**—passage 7—counting: 1117.6/ml.
**7 February 2017**—the paclitaxel experiment; preparation:

2 wells: 3 μl and 1 μl (1.5 ml of medium with cells).

**8 February 2017**—the paclitaxel experiment; observation.
**9 February 2017**—staining with Hoechst/PI (in the dark):

1. Add 3.75 µl of Hoechst 33342 resolved in PBS 1:1000 (999 µl PBS + 1 µl Hoechst) to a 12-well petri dish, 1.5 ml of medium per well.
2. Mix gently.
3. Incubate in an incubator at 37 °C for 25 min.
4. Add 3 µl of Pl resolved in PBS 1:1000 (999 µl PBS + 1 µl Hoechst) to a 12-well petri dish, 1.5 ml of medium per well.
5. Mix gently.
6. Incubate in an incubator at 37 °C for 5 min.
7. Observe for 2–3 h after staining.

**10 February 2017**—passage 10—counting: 934.42/ml.

**13 February 2017**—passage 11—counting: 466/ml.

**15 February 2017**—passage 11—changing the medium; passage 9—freezing.

**17 February 2017**—passage 12—counting: 1453125/ml.

**24 February 2017**—passage 13—counting: 1203125/ml.

**27 February 2017**—passage 14—counting: 1203125/ml, passage 11—changing the medium, passage 10—passaging to 11/12.

**28 February 2017**—preparing passage 13 for staining.

**3 March 2017**—passage 10—freezing, passage 15—counting: 1468750/ml.

**6 March 2017**—passage 16—counting: 2125000/ml.

**8 March 2017**—preparing BrdU specimens.

**9 March 2017**—preparing 53BP1 specimens.

**10 March 2017**—passage 17—counting: 812500/ml.

**13 March 2017**—passage 18—counting: 1390625/ml.

**15 March 2017**—passage 19—counting: 1031250/ml.

**20 March 2017**—passage 20—counting: 1640625/ml.

**22 March 2017**—passage 21—counting: 1890625/ml.

**24 March 2017**—passage 22—counting: 1671875/ml.

**27 March 2017**—passage 23—counting: 2890625/ml.

**28 March 2017**—rotenone administration: 0.5 µl, 1 µl, 2 µl, and 2.5 µl per well.

**29 March 2017**—staining with Hoechst.

**30 March 2017**—passage 24—counting: 906250/ml.

**6 April 2017**—passage 25—counting: 2296875/ml.

**10 April 2017**—passage 26—counting: 1484375/ml.

**18 April 2017**—passage 27—counting: 3015625/ml.

**24 April 2017**—passage 30—counting: 1578125/ml.

**28 April 2017**—passage 31—counting: 2390625/ml.
**2 May 2017**—passage 32—counting: 1656250/ml.
**4 May 2017**—passage 33—counting: 1656250/ml.
**8 May 2017**—passage 34—counting: 1937300/ml.
**12 May 2017**—passage 35—counting: 1578125/ml.
**15 May 2017**—passage 36—counting: 781250/ml; freezing.
**16 May 2017**—B-gal staining.
**17 May 2017**—B-gal—observations.
**31 May 2017**—defrosting passage 36, counting: 203125/ml.
**22 June 2017**—test for mycoplasma, PCR mix:

1 Isoamyl alcohol of DNTP
2 µl of starter
µl of buffer
1 µl of polymerase
25 µl of water

**26 June 2017**—passage 37—counting: 406250/ml.
**30 June 2017**—passage 38—counting: 390625/ml.
**3 July 2017**—passage 39—counting: 562500/ml.
**10 July 2017**—passage 40—counting: 437500/ml.
**25 July 2017**—electrophoresis (mycoplasma test).

# Appendix 2: Laboratory Notebook: *The Last Supper*

**22 May 2016**—precipitation—isolation of RNA:

1. Centrifuge for 10 min.
2. Discard supernatant.
3. Add 800 of TRIzol; resuspend the pellet.
4. Move the solution to 2 test tubes, 400 µl per each.
5. Add 80 µl of chloroform and shake strongly.
6. Prepare the mix: Phenol, Chloroform, Isoamyl alcohol, 200 µl.
7. Centrifuge for 10 min.
8. Phenol, Chloroform, Isoamyl alcohol, 200 µl.
9. Centrifuge for 5 min.
10. Precipitation: 200 µl of 96% ethanol, 70 µl of sodium acetate.
11. Centrifuge for 15 min.
12. Wash with 70% ethanol.
13. Leave to dry.

**30 May 2016**—PCR—cloning pyruvate kinase.
**31 May 2016**—PCR repetition.
**1 June 2016**—matrix cleaning:

1. 1:1—reaction mixture and Binding Buffer.
2. Move 800 µl of the solution to a Gene Jet purification column.



K. Żyniewicz, *The Body as Matter in Art and Science*,
https://doi.org/10.1007/978-3-032-13665-7

3. Add 700 µl of Wash Buffer, centrifuge for 30–60 s, and discard supernatant.
4. Centrifuge the empty column briefly.
5. Move the column to a 1.5 ml tube and centrifuge for 1 min.
6. Add 50 µl of Elution Buffer to the column and centrifuge for 1 min.
7. Store the cleaned DNA at −20 °C.

2 June 2016—controlling electrophoresis

**20 June 2016**—PCR:

Mix: 40 µl of Buffer, 40 µl of Enhancer, 2 µl of RT, 4 µl of dNTP, 4 µl of polymerase, and 53 µl of $H_2O$/10 µl per test tube.

**23 June 2016**—electrophoresis (10 µl of product per well):

PCR mix: 20 µl of buffer, 2.5 µl of dNTP, 2 µl of primers, 2 µl of polymerase, 73.5 µl of mQ, and 1 µl (our RT or control RT).

**29 June 2016**—reverse transcription:

1. Make a dilution of primers: 1 µl of primer, 39 µl of water.
2. Move 1 µl of the dilution to a PCR tube.
3. Add 5 µl of RNA.
4. Add 1 µl of DNTP.
5. Add 3 µl of water.
6. Incubate in a thermocycler for 5 min.
7. Add 4 µl of 5× First-Strand Buffer.
8. Add 1 µl of RNAsa OUT.
9. Add 1 µl of MDTT.
10. Add 1 µl of SuperScript III RT.

**13 July 2016**—electrophoresis.

**16 August 2016**—repetition of RNA isolation.

**18 August 2016**—repetition of reverse transcription.

**24 August 2016**—PCR on a mix of primers:

The mix of primers: 1 µl of PYK1, PYK2 + 18 µl water

PCR mix: 26 µl of Q5 Buffer, 3 µl of polymerase Q5, 3 µl of the mix of primers, 2.6 µl of DNTP, and 95.4 µl of water.

**26 August 2016**—PCR:

Mix: 26 µl of Q5 Buffer, 26 µl of Enhancer, 3 µl of the mix of primers, 2.6 µl of DNTP, and 66 µl of water/60 µl per well.

**6 October 2016**—PCR:

I mix: 20 μl of Buffer, 20 μl of Enhancer, 2 μl of DNTP, 1 μl of CDNA, 2 μl of enzyme, and 55 μl of water.
II mix: the same, but 75 μl of water
20 μl of product per well.

**10 October 2016**—checking electrophoresis: 6 μl of product per well.
**17 November 2016**—PCR:

Mix: 20 μl of Buffer, 20 μl of Enhancer, 2 μl of DNTP, 1 μl of the mix of primers, 2 μl of enzyme, and 55 μl of water.

**23 November 2016**—gradient: 40 μl of Buffer, 40 μl of Enhancer, 4 μl of DNTP, 2 μl of primers, 4 μl of enzyme, and 110 μl of water.
**29 November 2016**—electrophoresis.
**2 December 2016**—PCR: 10 μl of Buffer, 10 μl of Enhancer, 1 μl of DNTP, 1 μl of primers, 3 μl of RT, 1 μl of enzyme, and 24 μl of water.
**5 December 2016**—electrophoresis.
**15 December 2016**—PCR:

20 μl of Buffer, 20 μl of Enhancer, 2 μl of DNTP, 1 μl of primers, 2 μl of enzyme, 4 μl of cDNA, and 51 μl of water.

**20 December 2016**—PCR:

20 μl of Buffer, 20 μl of Enhancer, 2 μl of DNTP, 1 μl of primers, 2 μl of enzyme, 1 μl of cDNA, and 54 μl of water.

**5 January 2017**—PCR:

20 μl of Buffer, 20 μl of Enhancer, 2 μl of DNTP, 2 μl of primers, 2 μl of enzyme, 4 μl of RT, and 51 μl of water.

**11 January 2017**—electrophoresis.
**24 April 2017**—electrophoresis—checking RNA.
**25 April 2017**—reverse transcription:

Stage I: 0.5 μl of primer, 5 μl of RNA, 1 μl of DNTP, and 3.5 μl of water

Stage II: 2 μl of Buffer, 4 μl of $MgCl_2$, 2 μl of DTT, 1 μl of RNaseOUT, and 1 μl of SuperScript IIIRT.

**4 May 2017**—electrophoresis.
**8 May 2017**—gradient 49 °C–58 °C:

Mix: 40 μl of Buffer, 40 μl of Enhancer, 4 μl of Q5 polymerase, 4 μl of dNTP, 4 μl of starters, and 98 μl of water.

**18 May 2017**—electrophoresis.
**20 June 2017**—PCR:

The mix of primers: 2 μl of F + 2 μl of R + 46 μl of water.
PCR mix: 20 μl of Buffer, 20 μl of Enhancer, 2 μl of dNTP, 2 μl of Q5 polymerase, 1 μl of cDNA, 2 μl of primers, and 55 μl of water.

**21 June 2017**—electrophoresis.
**26 June 2017**—electrophoresis; checking of my cDNA.
**5 July 2017**—DNA purification (again).
**10 July 2017**—plasmid purification:

1. Resuspend the pelleted cells in 250 μl of the Resuspension Solution. Transfer the cell suspension to a microcentrifuge tube. The bacteria should be resuspended completely by vortexing or pipetting.
2. Add 250 μl of the Lysis Solution and mix thoroughly by inverting the tube 4–6 times until the solution becomes viscous and slightly clear.
3. Add 350 μl of the Neutralization Solution and mix immediately and thoroughly by inverting the tube 4–5 times.
4. Centrifuge for 5 min.
5. Transfer supernatant to a GeneJet spin column by decanting or pipetting.
6. Centrifuge for 1 min and discard the flow-through and place the column back into the same collection tube.
7. Add 500 μl of the Wash Solution to the spin column. Centrifuge for 30–60s and discard the flow-through. Place the column back into the same collection tube. (2×)
8. Discard the flow-through and centrifuge for 1 min to remove the residual solution.

9. Transfer the spin column into a fresh 1.5 ml microcentrifuge tube. Add 50 µl of the Elution Buffer to the centre of the GeneJet spin column membrane to elute the plasmid DNA. Take care not to contact the membrane with the pipet tip. Incubate at room temp. for 2 min and centrifuge for 2 min.
10. Discard the column and store the purified plasmid DNA at −20 °C.

**11 July 2017**—ligasis: the whole PCR product + 1 µl of the plasmid, 3 µl of ligase and 3 µl of ATP.

**13 July 2017**—cultivating the bacteria with the plasmid:

1. Add 20 µl of the plasmid after ligation to the chemo competent bacteria.
2. Incubate for 15 min.
3. Add 1 ml of SOB and incubate at 37 °C for 1 h.
4. Centrifuge for 1 min.
5. Discard supernatant.
6. Add 100 µl of water.
7. Spread the antibiotic on the surface of the solid bacteria medium.
8. Leave the petri dish to dry for 5 min.
9. Spread the bacteria on the surface of the medium.

**14 July 2017**—liquid culture:

1. Add 50 µl of ampicillin to the L medium (close to a fire).
2. Prepare 8 glass tubes and transfer 4 ml of culture to each of them.
3. Transfer one colony from the petri dish to the glass tube (using the pipet tip).
4. Incubate at 37 °C.

**17 July 2017**—plasmid purification.

**1 August 2017**—plasmid transformation:

1. Add 4 µl of the plasmid to the bacteria.
2. Keep the test tube on ice for 10 min.
3. Keep the test tube in hermoblock for 1.5 min; add 1 ml of SOB.
4. Incubate at 37 °C for 20 min.
5. Spread the antibiotic on the surface of the medium.
6. Spread the bacteria on the surface of the medium.

**4 August 2017**—liquid culture.
**9 August 2017**—PCR:

1 µl of the plasmid, 15 µl of insert, 10 µl of Buffer, 10 µl of Enhancer, 1 µl of dNTP, 1 µl of polymerase, and 12 µl of water.

**14 August 2017**—bacteria transformation.
**16 August 2017**—gradient—each PCR tube has a different annealing temperature:

Mix: 80 µl of Buffer, 80 µl of Enhancer, 8 µl of dNTP, 8 µl of primers, 8 µl of matrix, 8 µl of polymerase, and 208 µl of water.

**21 August 2017**—gradient:

Mix: 2 µl of plasmid, 20 µl of insert, 20 µl of Buffer, 20 µl of Enhancer, 2 µl of dNTP, 2 µl of polymerase, and 44 µl of water.

**25 August 2017**—bacteria transformation.
**4 September 2017**—bacteria transformation.
**11 September 2017**—plasmid purification.
**13 September 2017**—yeast culturing: 500 ml of YP medium, 5 ml of glucose, and 100 µl of ampicillin.
Yeast transformation:

1. Incubate the flask in the shaking incubator at 30 °C for around 4 h.
2. Denaturate a 1 ml sample of carrier DNA in a boiling water bath for 5 min and chill immediately in an ice/water bath.
3. Harvest the cells by centrifugation for 5 min, wash the pellet 2× in 2 ml of sterile water, and resuspend the pellet in 1.5 ml of sterile water.
4. Transfer the cells to a 1.5 ml microcentrifuge tube, centrifuge for 30 s, and discard supernatant.
5. Resuspend the pellet in 1.5 ml of sterile water and transfer to a fresh tube, one for each transformation. Centrifuge at top speed for 30 s and remove supernatant.
6. Prepare Tmix for all the transformations plus one extra. Keep it in ice water.

7. Add 360 μl of Tmix to each transformation tube and resuspend the cells by vortexing.
8. Incubate in a water bath at 42 °C for 35 min.
9. Centrifuge for 30 s and remove Tmix.
10. Add 1 ml of sterile water and resuspend the cells by vortexing.
11. Spread each T dilution to a separate plate.
12. Incubate at 30 °C for 3–4 days, count the number of transformants.

**23 October 2017**—liquid culture of transformants.

**24 October 2017**—brewer's yeast transformation.

**27 November 2017**—liquid culture of transformed brewer's yeast, reduction culture of brewer's yeast; and solid culture of lab strain W303.

**22 November 2017**—the second reduction culture of brewer's yeast; protein extraction:

1. Centrifuge 1 ml of liquid culture.
2. Resuspend the pellet in 100 μl of water and add 100 μl of NaOH (0.2 M).
3. Incubate at room temp.
4. Centrifuge for 1 min.
5. Resuspend the pellet with 50 μl of SDS Sample Buffer.
6. Put the sample to thermoblock at 100 °C for 3 min.

**29 November 2017**—yeast passage.

**1 December 2017**—checking of the plasmid in the bacteria:

1. Centrifuge 1 ml of the liquid culture.
2. Resuspend the pellet in 100 μl of Buffer.
3. Transfer the tube to thermoblock at 100 °C 5–10 min.
4. Centrifuge and remove supernatant.
5. Defrost the bacteria and add 10 μl of supernatant.
6. Keep the tube on ice for 15 min.
7. Keep in thermoblock at 42 °C for 1.5 min.
8. Keep the tube on ice for 5 min.
9. Add 1 ml of SOB.
10. Incubate at 37 °C for 30 min.
11. Centrifuge, resuspend the pellet, and transfer on the plate.
12. Incubate at 37 °C.

**4 December 2017**—Western blot—reincubation:

1. Wash with TBS.
2. Mark with Ponso for 5 min.
3. Incubate with milk for 1 h.
4. Incubation with I-row antibody (2 µl for 20 µl of milk) at 4 °C overnight.
5. Wash 3× for with TBS for 10 min.
6. Incubate with II-row antibody for 60 min.
7. Wash 3× with TBS for 10 min.

**7 November 2017**—control PCR:

The primer mix: 1 µl of F, 1 µl of R, ad 48 µl of water
PCR mix: 10 µl of Buffer, 10 µl of Enhancer, 1 µl of dNTP, 1 µl of primers, 1 µl of matrix (W303),1 µl of polymerase, and 26 µl of water.

**8 December 2017**—bacteria transformation.
**11 December 2017**—PCR—plasmid detection in strain W303; plasmid isolation from the bacteria:

PCR mix: 10 µl of Buffer, 10 µl of Enhancer, 1 µl of dNTP, 1 µl of primers, 1 µl of the bacteria, 26 µl of water.

**13 December 2017**—preparing of yeast strains for transformation with the plasmid: strains: NB80, Cw252, BMA 38a, BMA 64L, BYW 334, W303.
**19 December 2017**—yeast transformation:

1. Centrifuge for 1 min.
2. Resuspend the yeast pellet in 50% PEG.
3. Mix it with 15 µl of pre-boiled salmon sperm DNA and 36 µl of 1 M LiAC solution in $H_2O$.
4. Add 1 µl of plasmid DNA and vortex.
5. Incubate at 30 °C for 30 min and 1000 rpm.
6. Apply the heat shock at 42 °C in a water bath for 20 min.
7. Centrifuge shortly and resuspend the pellet in sterile $ddH_2O$.
8. Spread the solution on the selective agar medium.

**28 December 2017**—bacteria transformation; culturing the bacteria with the plasmid.

Transformation:

1. Centrifuge at 4 °C for 10 min.
2. Wash the pellet with water and centrifuge again at 4 °C for 10 min.
3. Add 400 µl of lithium acetate to the pellet, transfer the solution to a test tube, and store in a freezer.

Culturing: L medium + 100 µl of ampicillin; add bacteria and incubate at 37 °C.

**8 January 2018**—preparation of yeast for the next transformation; strains: W303, CW, NB, and BMA.

**9 January 2018**—plasmid purification; yeast transformation.

**26 January 2018**—bacteria transformation: (a) plasmid with an insert and a promotor; (b) plasmid with an insert alone.

**29 January 2018**—plasmids purification.

**6 February 2018**—yeast transformation.

**9 February 2018**—isolation of the plasmid from yeast:

Method I:

1. Centrifuge 5 ml.
2. Add 1 ml of Buffer/sorbitol and centrifuge.
3. Add 1 m of Buffer and 50 µl of enzyme, incubate at 37 °C for 30 min, and centrifuge at 3000 rpm.
4. Centrifuge at 6000 rpm for 1 min.
5. Discard supernatant.
6. Use the protocol from the isolation kit.

Method II:

1. Centrifuge 1 ml to obtain pellet.
2. Resuspend the pellet in 250 µl of Buffer.
3. Add a spoonful of glass beads and vortex for 3 min.
4. Use the protocol from the isolation kit.

Transformation of the bacteria with the isolated plasmid:

1. Add 15 μl of the plasmid to the bacteria.
2. Incubate on ice for 20 min.
3. Heat shock at 42 °C for 1.5 min.
4. Chill on ice for 1 min.
5. Add 1 ml of SOB
6. Incubate at 37 °C for 30 min.
7. Centrifuge, resuspend the pellet in water, and spread on the medium.

**13 September 2018**—control PCR and control digestion:

PCR mix: 10 μl of GC Buffer, 1 μl of dNTP, 1 μl of primers (diluted: 2 μl of 1 + 2 μl of 2 + 46 μl of water), 1 μl of polymerase, 1 μl of the plasmid (diluted 10×), and 36 μl of water

Digestion:

1. Plasmid 1—6 μl, 1 μl of polymerase, 1 μl of buffer, 2 μl of water
2. Plasmid 2—6 μl, 1 μl of polymerase, 1 μl of buffer, 2 μl of water

**14 February 2018**—control digestion:

10 μl of plasmid (pYCplac111), 1.5 μl of EcoR1, 1.5 μl of Pst1, 1.5 μl of orange buffer, and 0.5 μl of water.

**15 February 2018**—controlling electrophoresis, another plasmid digestion: 12 μl of the plasmid, 4 μl of water, 2 μl of polymerase, and 2 μl of buffer.

**1 March 2018**—transformation of bacteria with two new plasmids: p416ADM and p426ADM.

**5 March 2018**—isolation of plasmids from bacteria/digestion with new enzymes: 1 μl of XH01, 2 μl of BAMH1, 2 μl of Tango buffer, and 5 μl of vector.

**6 March 2018**—control PCR/plasmids digestion:

PCR mix: 10 μl of GC Buffer, 1 μl of dNTP, 1 μl of primers, 1 μl of polymerase fusion, 1 μl of matrix, and 36 μl of water.

**8 March 2018**—insert digestion: 6 µl of XH01, 12 µl of BAMH1, 12 µl of Tango buffer, and 30 µl of insert, incubation overnight.

**9 March 2018**—ligation, 2× for each of the plasmids:

1 µl of plasmid, 1 µl of insert, 1 µl of buffer for ligase, 1 µl of ATP, 1 µl of ligase, and 5 µl of water.

**12 March 2018**—bacteria transformation.

**14 March 2018**—plasmid isolation and digestion.

**15 March 2018**—control electrophoresis: digested and non-digested plasmids.

**20 March 2018**—transformation of yeast strain NB80:

Mix: 3350 µl of PEG,150 µl of ssDNA, 360 µl of LiAC, and water—440 µl /10 = 355 µl + 5 µl of plasmid.

**26 March 2018**—protein isolation:

1. Centrifuge 2 ml of culture.
2. Resuspend the pellet in 200 µl of 2 M LiAC.
3. Centrifuge and discard supernatant.
4. Resuspend the pellet in 200 µl of 0.4 M NaOH and keep on ice for 5 min.
5. Centrifuge and resuspend the pellet in 100 µl of SDS-PAGE.
6. Boil for 5 min.
7. Centrifuge, collect supernatant, and store at −80 °C.

**30 March 2018**—control Western blot.

**24 April 2018**—freeze-drying:

1. Prepare the solution: 2.5 g of milk and 0.25 g of monosodium glutamate.
2. Resuspend the yeast pellet in the solution.

**25 April 2018**—pre-fermentation.

**4 June 2018**—isolation of DNA from the fermenting brew:

1. Transfer 1.5 ml of liquid culture into a microcentrifuge tube.
2. Centrifuge for 1–5 min.

3. Add 200 µl of Harju Buffer.
4. Immerse the tube in a dry ice-ethanol bath for 2 min.
5. Transfer to a 95 °C water bath for 1 min.
6. Repeat the two previous steps.
7. Vortex for 30 s.
8. Add 200 µl of chloroform and vortex for 2 min.
9. Centrifuge at room temp. for 3 min.
10. Transfer the upper aqueous phase to a microcentrifuge tube containing 400 µl of ice-cold 100% ethanol. Mix by inversion or gentle vortexing.
11. Incubate at room temp. for 10 min.
12. Centrifuge at room temp. for 5 min.
13. Discard supernatant and wash the pellet with 0.5 ml of 70% ethanol.
14. Centrifuge at room temp. for 5 min.
15. Discard supernatant.
16. Air-dry the pellet at room temp. for 5 min.
17. Resuspend the pellet in 25–50 µl of TE or water.

**6 June 2018**—Western blot.
**11 June 2018**—PCR.

# Glossary

**53BP1** a protein involved in DNA repair in human cells. Its levels are lower in some types of cancer cells.

**adherent cells** cells that grow in a layer or multiple layers on a surface (as opposed to floating freely in the growth medium).

**apoptosis** otherwise known as programmed cell death. A controlled process that causes a damaged, mutated or senescent cell to "commit suicide" in a manner that does not cause serious damage to the surrounding tissue. It is also involved in shaping the organs and anatomic features during embryonic development.

**artifact** an unintended alteration or imperfection of results in experimental sciences; may be caused by errors or be inherent to some experimental procedure.

**autofluorescence** visible radiation emitted by all cells during fluorescent microscopy, regardless of presence of a dedicated fluorophore.

**β-gal** beta-galactosidase, an enzyme that breaks down sugars, such as galactose. It can also be used as a reporter gene, by decomposing X-gal into easily detectable, intensely blue product.

**bioptat** a tissue fragment collected by biopsy. It can be used for both diagnostic and research purposes.

**B lymphocytes** a subtype of white blood cells and crucial element of the immune system. Their main roles include antibody production and activation of other immune cells.



K. Żyniewicz, *The Body as Matter in Art and Science*,
https://doi.org/10.1007/978-3-032-13665-7

**BrdU** a molecule structurally similar to thymidine, one of the "building blocks" of DNA. It can replace thymine during replication, when new DNA strands are being synthesized. Thus, it can be used for detecting cell proliferation (each cellular division requires DNA replication).

**cell cultivating** a process of growing cells in laboratory setting. Requires a proper medium, nutrients, external conditions (such as temperature, $CO_2$ concentration, humidity etc.).

**centrifuging** a process of spinning samples at high velocities. It produces a strong outward force, which can divide molecules by their relative densities or separate soluble and insoluble substances that make up a sample.

**clumps** groups of cells "sticking" together. A phenomenon can be observed in cell cultures, where cells float freely in growth medium. It can sometimes be caused free DNA and cell debris from cells that have broken down.

**confocal microscope** a microscope which produces multiple two-dimensional "scans" of the sample at different depths. Can subsequently be used to create a three-dimensional model.

**DMEM (Dulbecco's modified Eagle's medium)** a medium for growing animal cells in vitro. It contains glucose and salts as well as essential amino acids and vitamins.

**DMSO** an organic solvent. In life sciences it is used as a cryoprotectant (among other things), preventing ice formation during cell freezing.

**electrophoresis** a process of separating biological molecules (DNA, RNA, proteins) by size. Negatively charged molecules migrate in an electric field, through a gel (agarose or acrylamide). Smaller molecules migrate faster. Molecules ofparticular size can subsequently be observed as discrete bands on the gel.

**Epstein-Barr virus** also known as herpes virus. Can be used to immortalize B lymphocytes.

**falcon** a plastic test tube with a conically shaped bottom and a twist-off cap. Comes in two sizes: 15 ml and 50 ml.

**flow cytometry** a technique for measuring properties of cells as well as for cell counting. It involves running a solution with suspended cells through a narrow capillary tube where a focused laser beam is used for detection. Ideally only a single cell should pass through the beam at any given time.

**growth curve** a method for visualizing the increase in number of cells (or increase in biomass) over time.

**Hayflick limit** the number of times that a given cell type can divide, before the divisions stop. Cancer cells and stem cells, as well as cells immortalized *in vitro*, are not subject to Hayflick limit.

**HeLa cells** cell line derived from human cervical cancer (named after the patient, Henrietta Lacks). One of the most commonly used cell lines in biological research.

**hypochlorite** a substance commonly found in bleaches. Can be used as a disinfectant, as it a strong oxidizing agent, quickly breaking down cellular structures both in microorganism and in animal cells.

**immortalizing** a process of altering cellular lines in a manner that allows them to overcome the Hayflick limit and proliferate indefinitely. This can be achieved by random mutagenesis, infecting cells with a virus, introducing a DNA construct expressing specific proteins or creating a hybrid with already immortalized cells.

**isopropanol** an organic compound, classified as alcohol. It is less toxic than ethanol. In biology, it can be used for DNA extraction.

**laminar flow cabinet** an enclosed bench for working with biological samples, designed to prevent microbiological contamination. Air is sucked in through a HEPA filter (which traps bacteria and viruses) and blown into the workspace, creating a slight overpressure that protects the samples from outside contamination.

**liquid nitrogen tank** a large vacuum flask for storing liquid nitrogen. It is also used for long-term storage of cell cultures.

**lysis** process of breaking down cellular membranes, necessary to extract molecules from a cell.

**mitochondria** a cellular compartment enclosed by a double membrane. Mitochondria are descended from a symbiotic bacterium, which was engulfed by a pre-eukaryotic. Mitochondria still contain their own (albeit considerably reduced) genome. They are involved in a number of crucial biological processes, including energy production, apoptosis and many others.

**mycoplasma** a genus of bacteria. Members of the genome are among the smallest bacteria identified to date. Some mycobacteria are known to cause diseases in humans. They are also known to contaminate cell cultures in laboratories. They are resistant to penicillin and its derivatives.

**non-adherent cells** cells that do not attach themselves to surfaces in *in vitro* cultures and float freely in the growth medium.

**overhang** a stretch of DNA on 5′ end of a primer, that is not complementary to the amplified sequence in PCR reaction. Overhangs can be used to add flanking regions to a PCR product (e.g. to introduce restriction enzyme sites).

**paclitaxel** a chemotherapeutic agent. It targets the proteins of cytoskeleton (a network of filaments presents in every cell, that gives it its shape and enables movement), disrupting cellular divisions.

**passage** a technique employed in cell cultivation. It involves removing a small volume of cell culture and diluting it in fresh medium. It enables cell cultivation for a prolonged period of time.

**Pasteur pipette** a small glass or plastic pipette. In cell culturing, it is often used in conjunction with a pump, to remove medium from adherent cell cultures.

**PBS** phosphate-buffered saline. A buffer commonly used in biological research. It's isotonic and non-toxic, making it a buffer of choice for working with cell cultures.

**PCR (Polymerase Chain Reaction)** a biochemical reaction for rapidly amplifying a selected DNA sequence. Aside from the enzyme that catalyses the reaction, reaction buffer and substrates, it also requires a template (e.g. genomic DNA) and primers. PCR reaction is very sensitive. Theoretically, it can produce billions of copies of selected DNA sequence from a single template molecule. One of the most fundamental and widely used techniques in molecular biology. It has applications in basic and applied research, medical diagnostics and forensics.

**pellet** solid phase that contains all the insoluble substances that gathered at the bottom of a test tube during centrifugation.

**penicillin** the first widely used antibiotic. Along with its derivative, ampicillin, it is widely used in biology—for example to protect cell cultures from bacteria.

**PFA (Paraformaldehyde)** a chemical substance used, among other things, for fixing cell cultures prior to microscopic observation.

**phalloidin** a fungus-derived toxin. Due to its strong binding to cytoskeleton, it is used to visualize it in fluorescent microscopy

**pipette** a laboratory tool for transferring small, measured volumes of liquid.

**pipettor** a laboratory tool used in conjunction with a glass or plastic graduated pipette. It contains a pump for sucking in or blowing out air.

**plasmid** a circular DNA molecule, capable of replication in bacteria and some other microorganisms. A commonly used tool for introducing new genetic elements into cells.

**primer** a short stretch of chemically synthesized, single-stranded DNA used in PCR reactions. Primers serve as "anchoring points" which mark the beginning and end points of amplified DNA sequence. They can also be used to add flanking sequences to the PCR products (see: "overhang").

**revko** a freezer which maintains a temperature of around −80 °C.

**rotenone** a substance that inhibits the electron transport chain (responsible for energy production) in mitochondria. Widely used as a pesticide.

**supernatant** liquid that contains all the soluble molecules that did not precipitate into the pellet during centrifugation.

**thermoblock** a device which consists of a plate with wells compatible in size with different types of test tubes. The plate can be heated or cooled to a desired temperature. Some thermoblocks can also agitate samples.

**Thoma chamber** a device for counting cells under a microscope. Consists of a microscope slide with a rectangular chamber and a laser-etched grid.

**T lymphocytes** a subtype of white blood cells and crucial element of the immune system. They can either kill a cancer cell or a virus-infected cell directly (by inducing apoptosis) or recruit other immune cells to respond to a perceived threat.

**trizol** a chemical reagent used to extract DNA, RNA or proteins from cells.

**Trypan blue** a dye used in microscopy to distinguish dead cells from living ones.

**trypsin** a protease (enzyme that breaks down proteins). Can be used to resuspend and harvest adherent cells.

**water bath** a laboratory instrument for incubating samples at a stable temperature for prolonged periods of time. Consists of a container filled with water and a heating system. Some water baths can also agitate incubated samples.

**Western blue** a chemical reagent. It is converted into a dark purple dye by alkaline phosphatase. Can be used for immunodetection of proteins by antibodies conjugated with alkaline phosphatase.

# Index




K. Żyniewicz, *The Body as Matter in Art and Science*,
https://doi.org/10.1007/978-3-032-13665-7